Building Big Data Applications

Building Big Data Applications

Krish Krishnan

ELSEVIER

ACADEMIC PRESS
An imprint of Elsevier

Academic Press is an imprint of Elsevier
125 London Wall, London EC2Y 5AS, United Kingdom
525 B Street, Suite 1650, San Diego, CA 92101, United States
50 Hampshire Street, 5th Floor, Cambridge, MA 02139, United States
The Boulevard, Langford Lane, Kidlington, Oxford OX5 1GB, United Kingdom

Notices
Knowledge and best practice in this field are constantly changing. As new research and experience broaden our understanding, changes in research methods, professional practices, or medical treatment may become necessary.

Practitioners and researchers must always rely on their own experience and knowledge in evaluating and using any information, methods, compounds, or experiments described herein. In using such information or methods they should be mindful of their own safety and the safety of others, including parties for whom they have a professional responsibility.

To the fullest extent of the law, neither the Publisher nor the authors, contributors, or editors, assume any liability for any injury and/or damage to persons or property as a matter of products liability, negligence or otherwise, or from any use or operation of any methods, products, instructions, or ideas contained in the material herein.

Library of Congress Cataloging-in-Publication Data
A catalog record for this book is available from the Library of Congress

British Library Cataloguing-in-Publication Data
A catalogue record for this book is available from the British Library

ISBN: 978-0-12-815746-6

For information on all Academic Press publications visit our website at
https://www.elsevier.com/books-and-journals

Publisher: Mara Conner
Acquisition Editor: Mara Conner
Editorial Project Manager: Joanna Collett
Production Project Manager: Punithavathy Govindaradjane
Cover Designer: Mark Rogers

Typeset by TNQ Technologies

Working together to grow libraries in developing countries

www.elsevier.com • www.bookaid.org

Dedicated to all my teachers

Contents

Preface

In the world that we live in today it is very easy to manifest and analyze data at any given instance. Space a very insightful analytics is worth every executive's time to make decisions that impact the organization today and tomorrow. Space this analytics is what we call Big Data analytics since the year 2010, and our teams have been struggling to understand how to integrate data with the right metadata and master data in order to produce a meaningful platform that can be used to produce these insightful analytics.

Not only is the commercial space interested in this we also have scientific research and engineering teams very much wanting to study the data and build applications on top off at. The effort's taken to produce Big Data applications have been sporadic when measured in terms of success why is that a question that is being asked by folks across the industry. In my experience of working in this specific space, what I have realized is that we are still working with data which is lost in terms of volumes come on and it is produced very fast on demand by any consumer leading to metadata integration issues. This metadata integration issue can be handled if we make it an enterprise solution, and all renters in the space need not necessarily worry about their integration with a Big Data platform. This integration is handled through integration tools that have been built for data integration and transformation. Another interesting perspective is that while the data is voluminous and it is produced very fast it can be integrated and harvested as any enterprise data segment. We require the new data architecture to be flexible, and scalable to accommodate new additions, updates, and integrations in order to be successful in building a foundation platform. This data architecture will differ from the third normal and star schema forms that we built the data warehouse from. The new architecture will require more integration and just in time additions which are more represented by NoSQL database architecture's and how architectures do. How do we get this go to success factor? And how do we make the enterprise realize that new approaches are needed to ensure success and accomplishing the tipping point on a successful implementation.

Our executives are always known for asking questions about the lineage of data and its traceability. These questions today can be handled in the data architecture and engineering provided we as an enterprise take a few minutes to step back and analyze why our past journeys journeys were not successful enough, and how we can be impactful in the future journey delivering the Big Data application. The hidden secret here is resting in the farm off governance within the enterprise. Governance, it is not about measuring people it is about ensuring that all processes have been followed and completed as requirements and that all specifics are in place for delivering on demand lineage and traceability.

In writing this book there are specific points that have been discussed about the architecture and governance required to ensure success in Big Data applications. The goal of the book is to share the secrets that have been leveraged by different segments of people in their big data application projects and the risks that they had to overcome to become successful.

The chapters in the book present different types of scenarios that we all encounter, and in this process the goals of reproducibility and repeatability for ensuring experimental success has been demonstrated. If you ever wondered what the foundational difference in building a Big Data application is the foundational difference is that the datasets can be harvested and an experimental stage can be repeated if all of the steps are documented and implemented as specified into requirements. Any team that wants to become successful in the new world needs to remember that we have to follow governance and implement governance in order to become measurable. Measuring process completion is mandatory to become successful and as you read it in the book revisit this point and draw the highlights from.

In developing this book there are several discussions that I have had with teams from both commercial enterprises as well as research organizations and thank all contributors for that time and insights and sharing the endeavors, it did take time to ensure that all the relevant people across these teams were sought out and tipping point of failure what discussed in order to understand the risks that could be identified and avoided in the journey. There are several reference points that has been added to chapters and while the book is not all encompassing by any means it does provide any team that wants to understand how to build a Big Data application choices of how success can be accomplished as well as case studies that vendors have shared showcasing how companies have implemented technologies to build the final solution.

I thank all vendors who provided material for the book and in particular IO-Tahoe, Teradata, and Kinetica for access to teams to discuss the case studies.

I thank my entire editorial and publishing team at Elsevier publishing for their continued support in this journey for their patience and support in ensuring completion of this book is what is in your hands today.

Last but not the least, I thank my wife and our two sons for the continued inspiration and motivation for me to write. Your love and support is a motivation.

Big Data introduction

This chapter will be a brief introduction to Big Data, providing readers the history, where are we today, and the future of data. The reader will get a refresher view of the topic.

   The world we live in today is flooded with data all around us, produced at rates that we have not experienced, and analyzed for usage at rates that we have heard as requirements before and now can fulfill the request. What is the phenomenon called as "Big Data" and how has it transformed our lives today? Let us take a look back at history, in 2001 when Doug Laney was working with Meta Group, he forecasted a trend that will create a new wave of innovation and articulated that the trend will be driven by the three V's namely volume, velocity, and variety of data. In the continuum in 2009, he wrote the first premise on how "Big Data" as the term was coined by him will impact the lives of all consumers using it. A more radical rush was seen in the industry with the embracement of Hadoop technology and followed by NoSQL technologies of different varieties, ultimately driving the evolution of new data visualization, analytics, storyboarding,and storytelling.

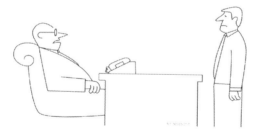

"The only Big Data letters I care about are the four Ms —— Make Me More Money!"

   In a lighter vein, SAP published a cartoon which read the four words that Big Data brings —"Make Me More Money"

   This is the confusion we need to steer clear of and be ready to understand how to monetize from Big Data.

   First to understand how to build applications with Big Data, we need to look at Big Data from both the technology and data perspectives.

Big Data delivers business value

The e-Commerce market has shaped businesses around the world into a competitive platform where we can sell and buy what we need based on costs, quality, and preference. The spread of services ranges from personal care, beauty, healthily eating, clothing,

Building Big Data Applications. https://doi.org/10.1016/B978-0-12-815746-6.00001-6

perfumes, watches, jewelry, medicine, travel, tours, investments, and the list goes on. All of this activity has resulted in data of various formats, sizes, languages, symbols, currencies, volumes, and additional metadata which we collectivity today call as "Big Data". The phenomenon has driven unprecedented value to business and can deliver insights like never before.

The business value did not and does not stop here; we are seeing the use of the same techniques of Big Data processing across insurance, healthcare, research, physics, cancer treatment, fraud analytics, manufacturing, retail, banking, mortgage, and more. The biggest question is how to realize the value repeatedly? What formula will bring success and value, how to monetize from the effort?

Take a step back for a moment and assess the same question with investments that has been made into a Salesforce or Unica or Endeca implementation and the business value that you can drive from the same. Chances are you will not have an accurate picture of the amount of return on investmentor the percentage of impact in terms of increased revenue or decreased spendor process optimization percentages from any such prior experiences. Not that your teams did not measure the impact, but they are unsure of expressing the actual benefit into quantified metrics. But in the case of a Big Data implementation, there are techniques to establish a quantified measurement strategy and associate the overall program with such cost benefits and process optimizations.

The interesting question to ask is what are organizations doing with Big Data? Are they collecting it, studying it, and working with it for advanced analytics? How exactly does the puzzle called Big Data fit into an organization's strategy and how does it enhance corporate decision-making?

To understand this picture better there are some key questions to think about and these are a few you can add more to this list:

- How many days does it take on an average to get answers to the question "why"?
- How many cycles of research does the organization do for understanding the market, competition, sales, employee performance, and customer satisfaction?
- Can your organization provide an executive dashboard along the ZachmanFramework model to provide insights and business answers on who, what, where, when, and how?
- Can we have a low code application that will be orchestrated with a workflow and can provide metrics and indicators on key processes?
- Do you have volumes of data but have no idea how to use it or do not collect it at all?
- Do you have issues with historical analysis?
- Do you experience issues with how to replay events? Simple or complex events?

The focus of answering these questions through the eyes of data is very essential and there is an abundance of data that any organization has today and there is a lot of hidden data or information in these nuggets that have to be harvested. Consider the following data:

- Traditional business systems—ERP, SCM, CRM, SFA
- Content management platforms
- Portals
- Websites
- Third-party agency data
- Data collected from social media
- Statistical data
- Research and competitive analysis data
- Point of sale data—retail or web channel
- Legal contracts
- Emails

If you observe a pattern here there is data about customers, products, services, sentiments, competition, compliance, and much more available. The question is does the organization leverage all the data that is listed here? And more important is the question, can you access all this data at relative ease and implement decisions? This is where the platforms and analytics of Big Data come into the picture within the enterprise. From the data nuggets that we have described 50% of them or more are internal systems and data producers that have been used for gathering data but not harnessing analytical value (the data here is structured, semistructured, and unstructured), the other 50% or less is the new data that is called Big Data (web data, machine data, and sensor data).

Big Data Applications are the answer to leveraging the analytics from complex events and getting the articulate insights for the enterprise. Consider the following example:

- Call center optimization—The worst fear of a customer is to deal with the call center. The fundamental frustration for the customer is the need to explain all the details about their transactions with the company they are calling, the current situation, and what they are expecting for a resolution, not once but many times (in most cases) to many people and maybe in more than one conversation. All of this frustration can be vented on their Facebook page or Twitter or a social media blog, causing multiple issues
  - They will have an influence in their personal network that will cause potential attrition of prospects and customers
  - Their frustration maybe shared by many others and eventually result in class action lawsuits
  - Their frustration will provide an opportunity for the competition to pursue and sway customers and prospects
  - All of these actions lead to one factor called as "revenue loss."If this company continues to persist with poor quality of service, eventually the losses will be large and even leading to closure of business and loss of brand reputation. It is

in situations like this where you can find a lot of knowledge in connecting the dots with data and create a powerful set of analytics to drive business transformation. Business transformation does not mean you need to change your operating model but rather it provides opportunities to create new service models created on data driven decisions and analytics.

The company that we are discussing here, let us assume, decides that the current solution needs an overhaul and the customer needs to be provided the best quality of service, it will need to have the following types of data ready for analysis and usage:

- Customer profile, lifetime value, transactional history, segmentation models, social profiles (if provided)
- Customer sentiments, survey feedback, call center interactions
- Product analytics
- Competitive research
- Contracts and agreements—customer specific

We should define a metadata-driven architecture to integrate the data for creating these analytics. There is a nuance of selecting the right technology and architecture for the physical deployment. A few days later the customer calls for support, the call center agent is now having a mash-up showing different types of analytics presented to them. The agent is able to ask the customer-guided questions on the current call and apprise them of the solutions and timelines, rather than ask for information; they are providing a knowledge service. In this situation the customer feels more privileged and even if there are issues with the service or product, the customer will not likely attrite. Furthermore, the same customer now can share positive feedback and report their satisfaction, thus creating a potential opportunity for more revenue. The agent feels more empowered and can start having conversations on cross-sell and up-sell opportunities. In this situation, there is a likelihood of additional revenue and diminished opportunities for loss of revenue. This is the type of business opportunities that Big Data analytics (internal and external) will bring to the organization, in addition to improving efficiencies, creating optimizations, and reducing risks and overall costs. There is some initial investment spent involved in creating this data strategy, architecture, and implementing additional technology solutions. The returnon investment will offset these costs and even save on license costs from technologies that may be retired post the new solution.

We see the absolute clarity that can be leveraged from an implementation of the Big Data—driven call center, which will provide the customer with confidence, the call center associate with clarity, the enterprise with fine details including competition, noise, campaigns, social media presence, the ability to see what customers in the same age group and location are sharing, similar calls, and results. All of this can be easily accomplished if we set the right strategy in motion for implementing Big Data applications. This requires us to understand the underlying infrastructure and how to leverage them for the implementation. This is the next segment of this chapter.

Healthcare example

In the past few years, a significant debate has emerged around healthcare and its costs. There are almost 80 million baby boomers approaching retirement, and economists forecast this trend will likely bankrupt Medicare and Medicaid in the near future. While healthcare reform and its new laws have ignited a number of important changes, the core issues are not resolved. It's critical we fix our system now, or else our $2.6 trillion in annual healthcare spending will grow to $4.6 trillion by 2020—one-fifth of our gross domestic product.

Data-rich and information-poor

Healthcare has always been datarich. Medicine has developed so quickly in the past 30 years that along with preventive and diagnostic developments, we have generated a lot of data: clinical trials, doctors' notes, patient therapies, pharmacists' notes, medical literature and, most importantly, structured analysis of the data sets in analytical models.

On the payer side, while insurance rates are skyrocketing, insurance companies are trying hard to vie for wallet share. However, you cannot ignore the strong influence of social media.

On the provider side, the small number of physicians and specialists available versus the growing need for them is becoming a larger problem. Additionally, obtaining second and third expert opinions for any situation to avoid medical malpractice lawsuits has created a need for sharing knowledge and seeking advice. At the same time, however, there are several laws being passed to protect patient privacy and data security.

On the therapy side, there are several smart machines capable of sending readings to multiple receivers, including doctors' mobile phones. We have become successful in reducing or eliminating latencies and have many treatment alternatives, but we do not know where best to apply them. Treatments that can work well for some, do not work well for others. We do not have statistics that can point to successful interventions, show which patients benefited from them, or predict how and where to apply them in a suggestion or recommendation to a physician.

There is a lot of data available, but not all of it is being harnessed into powerful information. Clearly, healthcare remains one of our nation's datarich, yet information-poor industries. It is clear that we must start producing better information, at a faster rate and on a larger scale.

Before cost reductions and meaningful improvements in outcomes can be delivered, relevant information is necessary. The challenge is that while the data is available today, the systems to harness it have not been available.

Big Data and healthcare

Big Data is information that is both traditionally available (doctors' notes, clinical trials, insurance claims data, and drug information), plus new data generated from social media, forums, and hosted sites (for example, WebMD) along with machine data. In healthcare, there are three characteristics of Big Data:

1. **Volume:** The data sizes are varied and range from megabytes to multiple terabytes
2. **Velocity:** The data production by machines, doctors' notes, nurses' notes, and clinical trials are all produced at different speeds and are highly unpredictable
3. **Variety:** The data is available or produced in a variety of formats but not all formats are based on similar standards

Over the past 5 years, there have been a number of technology innovations to handle Web 2.0-based data environments, including Hadoop, NoSQL, data warehouse appliances (iteration 3.0 and more), and columnar databases. There are several analytical models that have become available and late last year the Apache Software Foundation released a collection of statistical algorithms called Mahout. With so many innovations, the potential is there to create a powerful information processing architecture that will address multiple issues that face data processing in healthcare today:

- Solving complexity
- Reducing latencies
- Agile analytics
- Scalable and available systems
- Usefulness (getting the right information to the right resource at the right time)
- Improving collaboration

Potential solutions

How can Big Data solutions fix healthcare? A prototype solution flow is shown here. While this is not a complete production system flow, there are several organizations working on such models in small and large environments (Fig. 1.1).

An integrated system can intelligently harness different types of data using architectures like those of Facebook or Amazon to create a scalable solution. Using a textual processing engine like FRT Textual ETL (extract, transform, load) enables small and medium enterprises to write business rules in English. The textual data, images, and video data can be processed using any of the open source foundation tools. Data output from all these integrated processors will produce a rich data set and also generate an enriched column-value pair output. We can use the output along with existing enterprise data warehouse (EDW) and analytical platforms to create a strong set of models utilizing analytical tools and leveraging Mahout algorithms.

Using metadata-based integration of data and creating different types of solutions—including evidence-based statistics, clinical trial versus clinical diagnosis types of insights, patient dashboards for disease state management based on machine output and so on—lets us generate information that is rich, auditable, and reliable. This information can be used to provide better care, reduce errors, and create more confidence in sharing data with physicians in a social media outlet, thus providing more

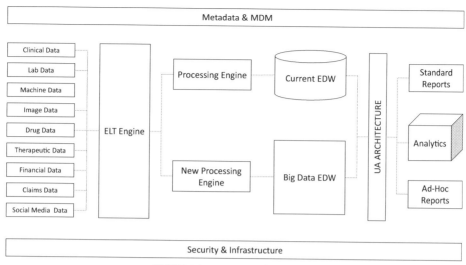

FIGURE 1.1 Prototype solution flow.

insights and opportunities. We can convert research notes from doctors that have been dormant into useable data, and create a global search database that will provide more collaboration and offer possibilities to share genomic therapy research.

When we can provide better cures and improve the quality of care, we can manage patient health in a more agile manner. Such a solution will be a huge step in reducing healthcare costs and fixing a broken system.

Eventually, this integrated data can also provide lineage into producing patient auditing systems based on insurance claims, Medicaid, and Medicare. It will also help isolate fraud, which can be a large revenue drain, and will create the ability to predict population-based spending based on disease information from each state. Additionally, integrated data will help drive metrics and goals to improve efficiency and ratios.

While all of these are lofty goals, Big Data-based solution approaches will help create a foundational step toward solving the healthcare crisis. There are several issues to confront in the data space, such as quality of data, governance, electronic health record (EHR) implementation, compliance, and safety and regulatory reporting. Following an open source type of approach, if a consortium can be formed to work with the U.S. Department of Health and Human Services, a lot of associated bureaucracy can be minimized. More vendor-led solution developments from the private and public sectors will help spur unified platforms that can be leveraged to create this blueprint.

Big Data Infrastructure is an interesting subject to discuss, as this forms the crux of how to implement Big Data applications. Let us take a step back and look at enterprise applications running across the organization.

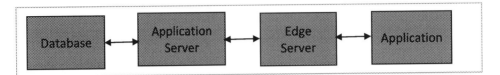

The traditional application processing happens when an application requests for either a read or write operation to the backend database. The request transits through the network often passing through an edge server to the application server and then to the database, and finally reverts back once the operation is complete. There are several factors to consider for improving and sustaining the performance which includes:

- Robust networks which can perform without any inhibition on throughput
- Fast performing edge servers that can manage thousands of users and queries
- Application servers with minimal interface delays and API's for performing the queries and operations
- Databases that are tuned for heavy transactional performance with high throughput.

All of these are very well-known issues when it comes to application performance and sustained maintenance of the same. The issue grows more complex when we need to use the data warehouse or a large database or an analytical model for these types of operations. The common issues that we need to solve include:

- Data reduction in dimensionality to accommodate the most needed and used attributes. This often results in multiple business intelligence projects that have a never-ending status.
- Data relationships management often becomes a burden or overload on the system.
- Delivering key analytics takes cycles to execute whether database or analytic model.
- Data lineage cannot be automatically traced.
- Data auditability is limited.
- Data aggregates cannot be drilled down or drilled across for all queries.

The issue is not with data alone, the core issue lies beneath the data layer, the infrastructure. The database is a phenomenal analytic resource and the schemas defined within the database are needed for all the queries and the associated multi-dimensional analytics. However, to load the schemas we need to define a fixed set of attributes from the dimensions as they are in source systems. These attributes are often gathered as business requirements, which is where we have a major missing point, the attributes are often defined by one business team and adding more attributes means issues, and we deliver too many database solutions and it becomes a nightmare. This is where we have created a major change with the

Big Data infrastructure which will be leveraged with applications. There are two platforms which we have created and they are Hadoop and NoSQL.

Hadoop—The platform originated in the world of Internet with Yahoo buying out Apache Nutch and implementing a platform that can perform infinite crawls of the web and provide search results. This infinite capability came with four basic design goals that were defined for Hadoop:

- System shall manage and heal itself
- Performance shall scale linearly
- Compute shall move to data
- Simple core, modular, and extensible

These goals were needed for the Internet because we do not have the patience to wait beyond a few milliseconds and often move away to other areas if we do not get answers. The biggest benefit of these goals is the availability of the platform $24 \times 7 \times 365$ with data always there as soon as it can be created and acquired into the platform. Today all the vendors have started adopting a Hadoop-driven interface and moving the on-premise to a cloud model and have integrated with in-memory processing and HDFS. We will see in upcoming chapters the details of the stack and how it has helped in multiple implementations.

Not-only-SQL (NoSQL) as we know it evolved into the web database platform that was designed to move away from the ACID compliant database and create a replication-based model to ingest and replicate data based on system requirements. We have seen the evolution of Cassandra, MongoDB, HBase, Amazon Dynamo, Apache Giraph, and MarkLogic. These NoSQL databases have all delivered solutions that have created analytics and insights like never before. These databases have been accepted into the enterprise but are yet to gain the adoption. We will discuss these databases and their implementations in the following chapters.

Building Big Data applications

Internet of Things evolves rapidly and grows at a fascinating pace bringing increasing opportunities to innovate at a continuum with capabilities to play and replay the events at occurrence and observe the effects as the event unfolds. Today we are equipped with the technology layers needed to make this paradigm shift and let the entire team of people whether in one location or across the globe to collaborate and understand the power of data. The paradigm shift did not occur easily and it took time to mature, but once it did hit reality the world has not stopped going through the tipping point multiple times.

The 10 commandments to building Big Data applications:

1. Data is the new paradigm shift. We need to understand that the world revolves around actions and counteractions from human beings and systems they are connected to. All of these actions produce data, which if harnessed and aligned will provide us a roadmap to what all happens in one action and its lifeline of activities from that point forward.

   a. Formats

   b. Geocodes

   c. Language

   d. Symbols

   e. Acquisition

   f. Ingestion

   g. Quality

2. Privacy—The issue of privacy is a very serious concern for all producers of data. Committees in the United States and European Union are working on defining privacy guidelines with respect to data in the new world. Evolution always occurs in stages, and we are now past the infancy stage of the evolution of the Internet of Things. Watch this space with interest as the data, its types, formats and details, and the devices evolve over the next decade.

3. Security—This area in the Internet of Things data lifecycle offers an interesting yes-and-no situation. The yes part is security requirements have been identified; the no part is these requirements have not been standardized into a set of regulations. This area is emerging rapidly with a great deal of the focus on isolating data, its transmission and encryption, its storage, and its lifecycle. Several articles on these topics are available that provide perspectives from the major stakeholders, and they all have solutions in their stack of offerings in regard to acquiring and managing data in the world of the Internet of Things.

4. Governance—In today's world, only a handful of companies across the globe have success in implementing a stellar data governance program. The worry here is that the remaining companies may have some aspects of a great data governance program but are hanging by thread in other critical areas. Based on my experience, I would say that the 30/70 rule applies to a data governance program's success/moderate success. The world of data for the Internet of Things needs more governance, more discipline and more analytics than ever, but, most important, it needs a highly managed lifecycle. If rapid resolutions are not achieved in this area and if it is not made a high priority, the journey for internal and Internet of Things data could be quite challenging.

5. Analytics in the world we live in surrounds us everywhere. We are getting more oriented to measure the value of everything we see and this is what the new world calls as "internet of things"and its driven analytics. We are in the need to develop an analytic ecosystem that can meet and exceed all the requirements of the new world of information.

6. Reporting is never going away from the enterprise, but can we get access to all the data and provide all the details needed in the reports? The answer for a new ecosystem has to be "yes". We need to provide a flexible and scalable reporting system for the enterprise. The goal is not around master data or enterprise data

but around acquiring all the raw data and then using that for discovery and finally reporting.

7. Artificial Intelligence is the new realm of intelligence that will be built for all enterprises. The intelligence is derived from both trained and untrained data sets. The artificial intelligence algorithms will be implemented across the entire data ecosystem, ranging from raw data to analytics databases. The algorithms can be opensource or enterprise or vendor provided. The implementation includes concepts including blockchain, robotic process automation, and lightweight data delivery systems.

8. Machine learning refers to an ecosystem of analytics and actions built on system outcomes from machines. These machines work 24/7/365 and can process data in continuum, which requires a series of algorithms, processes, code, analytics, action-driven outcomes, and no human interference. Work taking place for more than 25 years in this area has led to outcomes such as IBM Watson; TensorFlow, an open source library for numeric computation; Bayesian networks; hidden Markov model (HMM) algorithms; and Decision theory and Utility theory models of Web 3.0 processing. This field is the advancement of artificial intelligence algorithms and has more research and advancement published by Apache Software Foundation, Google, IBM, and many universities.

9. Smart everything
   a. Smart thermostats—The arrival of smart thermostats represents a very exciting and powerful Internet of Things technology. For example, based on the choices you make for controlling temperature, lighting, and timing inside your home, you can use your smartphone or tablet to control these home environment conditions from anywhere in the world. This capability has created much excitement in the consumer market. Millions of homes now have these devices installed. But what about the data part of this solution? To be able to do this smart thermostat magic, the device needs to be permanently connected to the Internet, not only to accommodate access, but more importantly to continuously send information to the power company or device manufacturer or both. Hence, the fear of the unknown: if anybody can get access to these devices and obtain your credentials from the stream of data, imagine what can happen next. Not only is identifying user preferences possible, someone hacking into the smart thermostat can monitor your presence in the home, break in when youarenot there or worse. Once someone has access to the network, theft of data can occur that possibly leads to other kinds of damage. Is this solution really that insecure? The answer is no. But ongoing work in the area of data governance and data privacy attempts to address the gaps in security that can cause concern. To help minimize these concerns, the underlying security of the data needs to be well managed.
   b. Smart cars—Electric automobiles manufactured by Tesla Motors and Nissan, for example, are touted for being purely electrically driven thanks to the

amount of computerization and logistics that make driving them an easy task. Similar smart car development efforts are underway with the Google driverless car experiments and testing and research at BMW, Mercedes Benz, and other auto manufacturers. All this smart car technology is fantastic and thought provoking, but smart cars have the capability to continuously communicate information—the condition of the vehicle and geographic coordinates of its location—to the manufacturer and possibly the dealer where the car was purchased. This capability can induce worry—more so over whether the transmission data is hack proof, for example, than whether the transmission is mechanically safe. And this concern is for good reason. If a transmission is intercepted, actions such as passing incorrect algorithms to the engine that may increase speed or cause a breakdown or an accident in a driverless vehicle are possible. Hacking into a smart car can also result in other disruptions such as changing the navigation system's map views. This fear of the unknown from smart car technology tends to be more with driverless cars than electric cars. Nevertheless, how can hacking smart cars be avoided? No set of regulations for this data and its security exist in the auto industry, and unfortunately rules are being created after the fact.

  c. Smart health monitoring—Remote monitoring of patients has become a new and advanced form of healthcare management. This solution benefits hospitals and healthcare providers, but it also creates additional problems for data management and privacy regulators. Monitored patients wear a smart device that is connected to the Internet so that the device can transmit data to a hospital, healthcare provider or third-party organization that provides data collection and on-call services for the hospital or provider. Although the data collected by a smart, wearable device generally is not specific to any single patient, enough data from these devices exists that can be hacked, for example, to obtain credentials for logging into the network. And once the network is compromised by a rogue login, consequences can be disastrous. For now, the situation with remote monitoring of patients is fairly well controlled, but security needs to be enhanced and upgraded for future implementations as the number of patients requiring remote monitoring increases. As demonstrated in the previous examples, electronic health record data requires enhanced management and governance.

10. Infrastructure for the enterprise will include Big Data platforms of Hadoop and NoSQL. The ecosystem design will not be successful without the new platforms and these platforms have provided extreme success in many industry situations.

Big Data applications—processing data

The processing of Big Data applications requires a step-by-step approach:

1. **Acquire data from all sources.** These sources include automobiles, devices, machines, mobile devices, networks, sensors, wearable devices, and anything that produces data.

2. **Ingest all the acquired data into a data swamp.** The key to the ingestion process is to tag the source of the data. Streaming data that needs to be ingested can be processed as streaming data and can also be saved as files. Ingestion also includes sensor and machine data.

3. **Discover data and perform initial analysis.** This process requires tagging and classifying the data based on its source, attributes, significance and need for analytics, and visualization.

4. **Create a data lake after data discovery is complete.** This process involves extracting the data from the swamp and enriching it with metadata, semantic data, and taxonomy and adding more quality to it as is feasible. This data is then ready to be used for operational analytics.

5. **Create data hubs for analytics.** This step can enrich the data with master data and other reference data, creating an ecosystem to integrate this data into the database, enterprise data warehouse, and analytical systems. The data at this stage is ready for deep analytics and visualization.

The key to note here is that steps 3, 4, and 5 are all helping in creating data lineage, data readiness with enrichment at each stage and a data availability index for usage.

Critical factors for success

While the steps for processing data are similar to what we do in the world of Big Data, the data here can be big, small, wide, fat, or thin and it can be ingested and qualified for usage. Several critical success factors will result from this journey:

- **Data:** You need to acquire, ingest, collect, discover, analyze and implement analytics on the data. This data needs to be defined and governed across the process. And you need to be able to handle more volume, velocity, variety, formats, availability, and ambiguity problems with data.

- **Business Goals:** The most critical success factor is defining business goals. Without the right goals, the data is neither useful, nor are the analytics and outcomes from the data useful.

- **Sponsors**: Executive sponsorship is needed for the new age of innovation to be successful. If no sponsorship is available, then the analytical outcomes, the lineage and linking of data, and the associated dashboards are all not happening and will be a pipe dream.

- **Subject Matter Experts:** The people and teams who are experts in the subject matter are needed to be involved in the Internet of Things journey; they are key to the success of the data analytics and using that analysis.

- **Sensor Data Analytics:** A new dimension of analytics is sensor data analytics. Sensor data is continuous and always streaming. It can be generated from an Apple iWatch, Samsung smartphone, Apple iPad, a smart wearable device, or a BMW i series, Tesla, or hybrid car. How do we monetize from this data? The answer is by implementing the appropriate sensor analytics programs. These programs require a team of subject and analytics experts to come together in a data science team approach for meeting the challenges and providing directions to the outcomes in the Internet of Things world. This move has started in many organizations but lacks direction and needs a chief analytics officer or chief data officer role to make it work in reality.

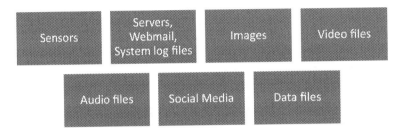

- **Machine Intelligence:** This success factor refers to an ecosystem of analytics and actions built on system outcomes from machines. These machines work 24/7/365 and can process data in continuum, which requires a series of algorithms, processes, code, analytics, action-driven outcomes, and no human interference. Work taking place for more than 25 years in this area has led to outcomes such as IBM Watson; TensorFlow, an open source library for numeric computation; Bayesian networks; hidden Markov model (HMM) algorithms; and Decision theory and Utility theory models of Web 3.0 processing. This field is the advancement of artificial intelligence algorithms and has more research and advancement published by Apache Software Foundation, Google, IBM and many universities.
- **Graph Databases:** In the world of the Internet of Things, graph databases represent the most valuable data processing infrastructure. This infrastructure exists because data will be streaming constantly and be processed by machines and people. It requires nodes of processing across infrastructure and algorithms with data captured, ingested, processed, and analyzed. Graph databases can scale up and out in these situations, and they can process with in-memory architectures such as Apache Spark, which provides a good platform for this new set of requirements.
- **Algorithms:** The algorithm success factor holds the keys to the castle in the world of the Internet of Things. Several algorithms are available, and they can be implemented across all layers of this ecosystem.

Risks and pitfalls

No success is possible without identifying associated risks and pitfalls. In the world driven by the Internet of Things, the risks and pitfalls are all similar to those we need to handle on a daily basis in the world of data. The key here is that, data volume can cause problems created by excessive growth and formats.

- Lack of data: A vital area to avoid within the risks and pitfalls is a lack of data, which is not identifying the data required in this world driven by the Internet of Things architecture. This pitfall can lead to disaster right from the start. Be sure to define and identify the data to collect and analyze, its governance and stewardship, its outcomes and processing—it is a big pitfall to avoid.
- Lack of governance: Data lacking governance can kill a program. No governance means no implementation, no required rigor to succeed, and no goals to be measured and monitored. Governance is a must for the program to succeed in the world of the Internet of Things.
- Lack of business goals: No key business actions or outcomes can happen when there are no business goals established. Defining business goals can provide clear direction on which data and analytics need to be derived with Internet of Things data and platforms. Two important requirements for these goals helps avoid this important pitfall: one is executive sponsorship and involvement, and the other is governance. Do not enter into this realm of innovative thinking and analytics without business goals.
- Lack of analytics: No analytics can lead to total failure and facilitates nonadoption and a loss of interest in the Internet of Things program. Business users need to be involved in the program and asked to define all the key analytics and business applications. This set of analytics and applications can be documented in a roadmap and delivered in an implementation plan. A lack of analytics needs to be avoided in all programs related to the Internet of Things.
- Lack of algorithms: No algorithms can create no results and translates to nonadoption of the program. A few hundred algorithms can be implemented across Internet of Things platforms and data. These algorithms need to be understood and defined for implementation, which requires some focus and talent in the organization both from a leadership and team perspective. Algorithms are expected to evolve over time and need to be defined in the roadmap.
- Incorrect applications: The use of incorrect applications tends to occur from business users with a lack of understanding of the data on the Internet of Things platform, and it is a pitfall to avoid early on. The correct applications can be defined as proof-of-value exercises and executed to provide clarity of the applications. Proof of value is a cost-effective solution architecture build out and scalability for the Internet of Things platform.

- Failure to govern: If no effective data governance team is in place, implementing, or attempting any data or analytics, can be extremely challenging. This subject has been a sore point to be resolved in all aspects of data but has not been implemented successfully very often. For any success in the Internet of Things, the failure to govern pitfall needs to be avoided with a strong and experienced data governance team in place.

Some of the best in class applications we have seen and experienced in this new land of opportunity are the National Institutes of Health (NIH)'s Precision Medicine Initiative (PMI), fraud analytics in healthcare, and financial analytics with advanced clustering and classification techniques on mobile infrastructure. More opportunities exist in terms of space exploration, smart cars and trucks, and new forays into energy research. And donot forget the smart wearable devices and devices for pet monitoring, remote communications, healthcare monitoring, sports training, and many other innovations.

Additional reading

Competitive Strategy: Techniques for Analyzing Industries and Competitors by Micheal Porter.

Keeping Up with the Quants: Your Guide to Understanding and Using Analytics by Thomas Davenport.

Own the A.I. Revolution: Unlock Your Artificial Intelligence Strategy to Disrupt Your Competition by Neil Sahota and Michael Ashley.

SuperCrunchers: Why Thinking-By-Numbers is the New Way To Be Smart by Ian Ayers.

TheNumerati by Stephen Baker.

They've Got Your Number...: Data, Digits and Destiny — how the Numerati are changing our Lives by Stephen Baker.

2

Infrastructure and technology

This chapter will introduce all the infrastructure components and technology vendors who are providing services. We will discuss in detail the components and their integration, the technology limitations if any to be known, specifics on the technology for users to identify and align with.

The first rule of any technology used in a business is that automation applied to an efficient operation will magnify the efficiency. The second is that automation applied to an inefficient operation will magnify the inefficiency.
Source: Brainy Quote—Bill Gates

Introduction

In the previous chapter we discussed the complexities associated with big data. There is a three-dimensional problem with processing this type of data; the dimensions being the volume of the data produced, the variety of formats, and the velocity of data generation. To handle any of these problems in traditional data processing architecture is not a feasible option. The problem by itself did not originate in the last decade and has been something that was being solved by various architects, researchers, and organizations over the years. A simplified approach to large data processing was to create distributed data processing architectures and manage the coordination by programming language techniques. This approach while solving the volume requirement did not have the capability to handle the other two dimensions. With the advent of Internet and search engines, the need to handle the complex and diverse data became a necessity and not a one-off requirement. It is during this time in the early 1990s a slew of distributed data processing papers and associated algorithms and techniques were published by Google, Stanford University, Dr.Stonebraker, Eric Brewer, Doug Cutting (Nutch Search Engine), and Yahoo among others.

Today the various architectures and papers that were contributed by these and other developers across the world have culminated into several open source projects under the Apache Software Foundation and the NoSQL movement. All of these technologies have been identified as big data processing platforms including Hadoop, Hive, HBase, Cassandra, and MapReduce. NoSQL platforms include MongoDB, Neo4J, Riak, Amazon DynamoDB, MemcachedDB, BerkleyDB, Voldemort, and many more. Though many of these platforms were originally developed and deployed for solving the data processing needs of web applications and search engines, they have been evolved to support other

Building Big Data Applications. https://doi.org/10.1016/B978-0-12-815746-6.00002-8

data processing requirements. In the rest of this chapter, the intent is to provide you with how data processing is managed by these platforms. This chapter is not a tutorial for step-by-step configuration and usage of these technologies. There are references provided at the end of this chapter for further reading.

Distributed data processing

Before we proceed to understand how big data technologies work and see associated reference architectures, let us take a recap at distributed data processing.

Distributed data processing has been in existence since late 1970s. The primary concept was to replicate the DBMS in a master—slave configuration and process data across multiple instances. Each slave would engage in a two-phase commit with its master in a query-processing situation. Several papers exist on the subject and how its early implementations have been designed and authored by Dr.Stonebraker, Teradata, UC Berkley Departments, and others.

Several commercial and early open source DBMS systems have addressed large-scale data processing with distributed data management algorithms; however, they all faced problems in the areas of concurrency, fault tolerance, supporting multiple copies of data, and distributed processing of programs. A bigger barrier was the cost of infrastructure (Fig. 2.1).

Why distributed data processing failed in the relational architecture? The answer to this question lies in multiple dimensions:

- Dependency on RDBMS
  - ACID compliance for transaction management
  - Complex architectures for consistency management
  - Latencies across the system
    - Slownetworks
    - RDBMS IO
    - SAN architecture
- Infrastructure cost
- Complex processing structure

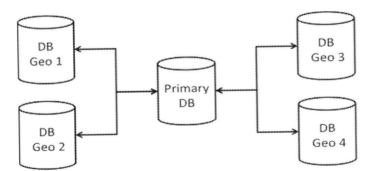

FIGURE 2.1 Distributed data processing in the relational database management system (RDBMS).

- Minimal fault tolerance within infrastructure and expensive fault tolerance solutions

Due to the inherent complexities and the economies of scale, the world of data warehousing did not adopt to the concept of large-scale distributed data processing. On the other hand the world of OLTP adopted and deployed distributed data processing architecture, using heterogeneous and proprietary techniques, though this was largely confined to large enterprises, where latencies were not the primary concern. The most popular implementation of this architecture is called as client–server data processing.

The client–server architecture had its own features and limitations, but it provided limited scalability and flexibility:

- Benefits
  - Centralization of administration, security, and setup
  - Back-up and recovery of data is inexpensive, as outages can occur at server or a client and can be restored
  - Scalability of infrastructure by adding more server capacity or client capacity can be accomplished. The scalability is not linear
  - Accessibility of server from heterogeneous platforms locally or remotely
  - Clients can use servers for different types of processing
- Limitations
  - Server is the central point of failure
  - Very limited scalability
  - Performance can degrade with network congestion
  - Too many clients accessing a single server cannot process data in a quick time

In the late1980s and early 1990s there were several attempts at distributed data processing in the OLTP world, with the emergence of "object oriented programming" and "object store databases". We learned that with effective programming and non-relational data stores, we could effectively scale up distributed data processing across multiple computers. It was at the same time the Internet was gaining adoption and web-commerce or e-commerce was beginning to take shape. To serve Internet users faster and better, several improvements rapidly emerged in the field of networking with higher speeds and bandwidth while lowering costs. At the same time the commoditization of infrastructure platforms reduced the cost barrier of hardware.

The perfect storm was created with the biggest challenges that were faced by web applications and search engines, which is unlimited scalability while maintaining sustained performance at the lowest computing cost. Though this problem existed prior to the advent of Internet, its intensity and complexity were not comparable to what web applications brought about. Another significant movement that was beginning to gain notice was nonrelational databases (specialty databases) and NoSQL (not only SQL), Combining the commoditization of infrastructure and distributed data processing techniques including NoSQL, highly scalable and flexible data processing architectures were designed and implemented for solving large-scale distributed processing by leading companies including Google, Yahoo, Facebook, and Amazon. The fundamental tenets that are common in this new architecture are the

- Extreme Parallel processing—ability to process data in parallel within a system and across multiple systems at the same time
- Minimal database usage—RDBMS or DBMS will not be the central engine in the processing, removing any architecture limitations from the database ACID compliance
- Distributed File based storage—data is stored in files, which is cheaper compared to storing on a database. Additionally data is distributed across systems, providing built-in redundancy
- Linearly scalable infrastructure—every piece of infrastructure added will create 100% scalability from CPU to storage and memory
- Programmable APIs—all modules of data processing will be driven by procedural programming APIs, which allows for parallel processing without the limitations imposed by concurrency. The same data can be processed across systems for different purposes or the same logic can process across different systems. There are different case studies on these techniques.
- High-speed replication—data is able to replicate at high speeds across the network
- Localized processing of data and storage of results—ability to process and store results locally, meaning compute and store occur in the same disk within the storage architecture. This means one needs to store replicated copies of data across disks to accomplish localized processing
- Fault tolerance—with extreme replication and distributed processing, system failures could be rebalanced with relative ease, as mandated by web users and applications (Fig. 2.2).

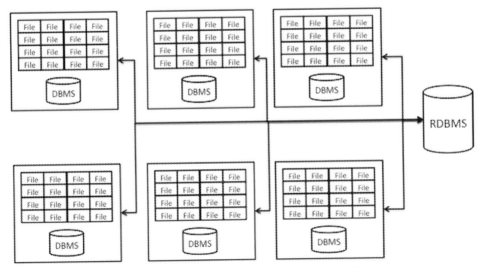

FIGURE 2.2 Generic new generation distributed data architecture.

With the features and capabilities discussed here, the limitations of distributed data processing with relational databases are not a real barrier anymore. The new generation architecture has created a scalable and extensible data processing environment for web applications and has been adopted widely by companies that use web platforms. Over the last decade many of these technologies have been committed back to open source community for further development by innovators across the world (refer to Apache foundation page for committers across projects). The new generation data processing platforms including Hadoop, Hive, HBase, Cassandra, MongoDB, Neo4J, DynamoDB, and more are all products of these exercises, which are discussed in this chapter.

There is a continuum of technology development in this direction (by the time we are finished with this book, there will be newer developments, that can be found on the website of this book).

Big data processing requirements

What is unique about big data processing? What makes it different or mandates new thinking? To understand this better let us look at the underlying requirements. We can classify big data requirements based on its characteristics

- Volume
    - Size of data to be processed is large; it needs to be broken into manageable chunks
    - Data needs to be processed in parallel across multiple systems
    - Data needs to be processed across several program modules simultaneously
    - Data needs to be processed once and processed to completion due to volumes
    - Data needs to be processed from any point of failure, since it is extremely large to restart the process from beginning
- Velocity
    - Data needs to be processed at streaming speeds during data collection
    - Data needs to be processed for multiple acquisition points
- Variety
    - Data of different formats needs to be processed
    - Data of different types needs to be processed
    - Data of different structures need to be processed
    - Data from different regions need to be processed
- Complexity
    - Big data complexity needs to use many algorithms to process data quickly and efficiently
    - Several types of data need multi-pass processing and scalability is extremely important

Technologies for big data processing

There are several technologies that have come and gone in the data processing world, from the mainframes, to two tier databases, to VSAM files. Several programming languages have evolved to solve the puzzle of high-speed data processing and have either stayed niche or never found adoption. After the initial hype and bust of the Internet bubble, there came a moment in the history of data processing that caused unrest in the industry, the scalability of the Internet search. Technology startups like Google, RankDex(now known as Baidu), and Yahoo, open source projects like Nutch were all figuring out how to increase the performance of the search query to scale infinitely. Out of these efforts came the technologies, which are now the foundation of big data processing.

MapReduce

MapReduce is a programming model for processing extremely large sets of data. Google originally developed it for solving the scalability of search computation. Its foundations are based on principles of parallel and distributed processing without any database dependency. The flexibility of MapReduce lies in the ability to process distributed computations on large amounts of data on clusters of commodity servers, with simple task based models for management of the same.

The key features of MapReduce that makes it the interface on Hadoop or Cassandra include the following:

- Automatic parallelization
- Automatic distribution
- Faulttolerance
- Status and monitoring tools
- Easy abstraction for programmers
- Programming language flexibility
- Extensibility

MapReduce programming model

MapReduce is based on functional programming models largely from Lisp. Typically the users will implement two functions:

- **Map (in_key, in_value) -> (out_key, intermediate_value) list**
    - Map function written by the user, will receive an input pair of keys and values, and postcomputation cycles produces a set of intermediate key/value pairs.
    - Library functions then are used to group together all intermediate values associated with an intermediate key I and passes them to the Reduce function.

- **Reduce (out_key, intermediate_value list) - > out_value list**
  - The Reduce function written by the user will accept an intermediate key I, and the set of values for the key.
  - It will merge together these values to form a possibly smaller set of values.
  - Reducer outputs are just zero or one output value per invocation.
  - The intermediate values are supplied to the reduce function via an iterator. The iterator function allows us to handle large lists of values that cannot fit in memory or a single pass.

MapReduce Google architecture

In the original architecture that Google proposed and implemented, MapReduce consisted of the architecture and components as described in Fig. 2.3. The key pieces of the architecture include the following:

- A GFS cluster
  - A single *master*
  - Multiple *chunkservers* (workers or slaves) per master
  - Accessed by multiple *clients*
  - Running on commodity Linux machines
- A file
  - Represented as fixed-sized *chunks*
  - Labeled with 64-bit unique global IDs
  - Stored at chunkservers and 3-way mirrored across chunkservers

In the GFS cluster, input data files are divided into chunks (64 MB is the standard chunk size), each assigned its unique 64-bit handle, and stored on local chunkserver systems as files. To ensure fault tolerance and scalability, each chunk is replicated at least once on another server, and the default design is to create three copies of a chunk (Fig. 2.4).

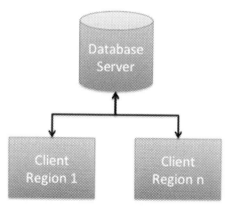

FIGURE 2.3 Client—server architecture.

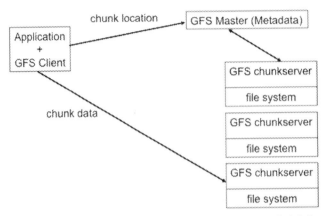

FIGURE 2.4 Google MapReduce cluster. *Image source—Google briefing.*

If there is only one master there is a potential bottleneck in the architecture right? The role of the master is to communicate to the clients: *chunkservers* have what chunks and their metadata information. Client's tasks then interact directly with *chunkservers* for all subsequent operations, and use the master only in a minimal fashion. The master therefore never becomes or is in a position to become the *bottleneck*.

Another important issue to understand in the GFS architecture is the single point of failure (SPOF) of the master node and all the metadata that keeps track of the chunks and their state. To avoid this situation, GFS was designed to have the master keep data in memory for speed, keep a log on the master's local disk, and replicate the disk across remote nodes. This way if there is a crash in the master node, a shadow can be up and running almost instantly.

The master stores three types of metadata:

- File and chunk names or *namespaces*
- Mapping from files to chunks, i.e., the chunks that make up each file
- Locations of each chunk's replicas—The replica locations for each chunk is stored on the local chunkserver apart from being replicated, and the information of the replications is provided to the master at startup or when a chunkserver is added to a cluster. Since the master controls the chunk placement it always updates meta-data as new chunks get written.

The master keeps track on the health of the entire cluster through handshaking with all the chunkservers. Periodic checksums are executed to keep track of any data corruption. Due to the volume and scale of processing, there are chances of data getting corrupt or stale.

To recover from any corruption, GFS appends data as it is available rather than update existing dataset, which provides the ability to recover from corruption or failure quickly. When a corruption is detected, with a combination of frequent checkpoints, snapshots, and replicas, data is recovered with minimal chance of data loss. The architecture results in data unavailability for a short period but not data corruption.

The GFS architecture has the following strengths:

- Availability
  - Triple replication-based redundancy (or more if you choose)
  - Chunk replication
  - Rapid failovers for any master failure
  - Automatic replication management
- Performance
  - The biggest workload for GFS is read on large data sets, which based on the architecture discussion, will be a nonissue.
  - There are minimal writes to the chunks directly, thus providing auto availability
- Management
  - GFS manages itself through multiple failure modes
  - Automatic load balancing
  - Storage management and pooling
  - Chunk management
  - Failover management
- Cost
  - Is not a constraint due to use of commodity hardware and Linux platforms

The platforms combined together along with proprietary techniques enabled Google and other companies that adopted the technologies and customized it further to enable performance within their organizations.

A pureplay architecture of MapReduce + GFS (or other similar filesystem) deployments can become messy to manage on large environments. Google has created multiple proprietary layers that cannot be adapted by any organization. In order to ensure management and deployment, the most extensible and successful platform for MapReduce is Hadoop, which we will discuss in later sections of this chapter. There are many variants of MapReduce programming today including SQL-MapReduce (AsterData), GreenplumMapReduce, MapReduce with Ruby, MongoDBMapReduce to name a few.

Hadoop

The most popular word in the industry at the time of writing this book, Hadoop has taken the world by storm in providing the solution architecture to solve big data processing on a cheaper commodity platform with faster scalability and parallel processing. This section's goal is to introduce you to Hadoop and cover the core components of Hadoop.

History

No book is complete without the history of Hadoop. The project started out as a sub-project in the open source search engine called Nutch, which was started by Mike Cafarella and Doug Cutting. In 2002 the two developers and architects realized that while they built a successful crawler, it cannot scale up or scale out. Around the same time, Google announced the availability of GFS to developers, which was quickly followed by the papers on MapReduce in 2002.

In 2004 the Nutch team developed the NDFS, an open source distributed filesystem, which was the open source implementation of GFS. The NDFS architecture solved the storage and associated scalability issues. In 2005, the Nutch team completed the port of Nutch algorithms to MapReduce. The new architecture would enable processing of large and unstructured data with unsurpassed scalability.

In 2006 the Nutch team of Cafarella and Cutting created a subproject under Apache Lucene and called it Hadoop (named after Doug Cutting's son's toy elephant). Yahoo adopted the project and in January 2008 released the first complete project release of Hadoop under open source.

The first generation of Hadoop consisted of HDFS (modeled after NDFS) distributed filesystem and MapReduce framework along with a coordinator interface and an interface to write and read from HDFS. When the first generation of Hadoop architecture was conceived and implemented in 2004 by Cutting and Cafarella, they were able to automate a lot of operations on crawling and indexing on search, and improved efficiencies and scalability. Within a few months they reached an architecture scalability of 20 nodes running Nutch without missing a heartbeat. This provided Yahoo the next move to hire Cutting and adopt Hadoop to become one of its core platforms. Yahoo kept the platform moving with its constant innovation and research. Soon many committers and volunteer developers/testers started contributing to the growth of a healthy ecosystem around Hadoop.

At this time of writing (2018), we have seen two leading distributors of Hadoop with management tools and professional services emerge—Cloudera and HortonWorks. We have also seen the emergence of Hadoop-based solutions from MapR, IBM, Teradata, Oracle, and Microsoft. Vertica, SAP, and others are also announcing their own solutions in multiple partnerships with other providers and distributors.

The most current list at Apache's website for Hadoop lists the top level stable projects and releases and also incubated projects which are evolving Fig. 2.5.

Hadoop core components

At the heart of the Hadoop framework or architecture there are components that can be called as the foundational core. These components include the following (Fig. 2.6):

Let us take a quick look at these components and further understand the ecosystem evolution and recent changes.

- **Hadoop Common**: The common utilities that support the other Hadoop modules.
- **Hadoop Distributed File System (HDFS™)**: A distributed file system that provides high-throughput access to application data.
- **Hadoop YARN**: A framework for job scheduling and cluster resource management.
- **Hadoop MapReduce**: A YARN-based system for parallel processing of large data sets.

Other Hadoop-related projects at Apache include:

- Ambari™: A web-based tool for provisioning, managing, and monitoring Apache Hadoop clusters which includes support for Hadoop HDFS, Hadoop MapReduce, Hive, HCatalog, HBase, ZooKeeper, Oozie, Pig and Sqoop. Ambari also provides a dashboard for viewing cluster health such as heatmaps and ability to view MapReduce, Pig and Hive applications visually alongwith features to diagnose their performance characteristics in a user-friendly manner.
- Avro™: A data serialization system.
- Cassandra™: A scalable multi-master database with no single points of failure.
- Chukwa™: A data collection system for managing large distributed systems.
- HBase™: A scalable, distributed database that supports structured data storage for large tables.
- Hive™: A data warehouse infrastructure that provides data summarization and ad hoc querying.
- Mahout™: A Scalable machine learning and data mining library.
- Pig™: A high-level data-flow language and execution framework for parallel computation.
- Spark™: A fast and general compute engine for Hadoop data. Spark provides a simple and expressive programming model that supports a wide range of applications, including ETL, machine learning, stream processing, and graph computation.
- Tez™: A generalized data-flow programming framework, built on Hadoop YARN, which provides a powerful and flexible engine to execute an arbitrary DAG of tasks to process data for both batch and interactive use-cases. Tez is being adopted by Hive™, Pig™ and other frameworks in the Hadoop ecosystem, and also by other commercial software (e.g. ETL tools), to replace Hadoop™ MapReduce as the underlying execution engine.
- ZooKeeper™: A high-performance coordination service for distributed applications.

FIGURE 2.5 Apache top level Hadoop projects.

HDFS

The biggest problem experienced by the early developers of large-scale data processing was the ability to break down the files across multiple systems and process each piece of the file independent of the other pieces, but yet consolidate the results together in a single result set. The secondary problem that remained unsolved for was the fault tolerance both at the file processing level and the overall system level in the distributed processing systems.

With GFS the problem of scalingout processing across multiple systems was largely solved. HDFS, which is derived from NDFS, was designed to solve the large distributed data processing problem. Some of the fundamental design principles of HDFS are the following:

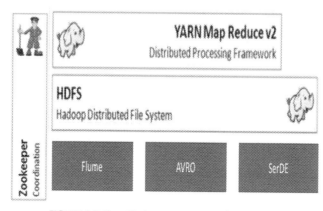

FIGURE 2.6 Core Hadoop components (circa 2017).

- Redundancy—hardware will be prone to failure and processes can run out of infra-structure resources. But redundancy built into the design can handle these situations
- Scalability—linear scalability at a storage layer is needed to utilize parallel process-ing at its optimum level. Designing for 100% linear scalability
- Fault tolerance—automatic ability to recover from failure
- Cross platform compatibility
- Compute and storage in one environment—data and computation colocated in the same architecture will remove a lot of redundant I/O and disk access

The three principle goals of HDFS are the following:

- Process extremely large files—ranging from multiple gigabytes to petabytes
- Streaming data processing—read data at high throughput rates and process data on read
- Capability to execute on commodity hardware—no special hardware requirements

These capabilities and goals form the robust platform for data processing that exists in the Hadoop platform today.

HDFS architecture

Fig. 2.7 shows the architecture of HDFS. The underlying architecture of HDFS represents master/slave architecture. The main components of HDFS are the following

- NameNode
- DataNode
- Image
- Journal

NameNode

The NameNode is a single master server that manages the filesystem namespace and regulates access to files by clients. Additionally the NameNode manages all the

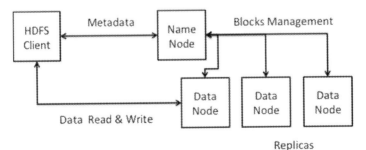

FIGURE 2.7 HDFS architecture.

operations like opening, closing, moving, naming, renaming of files, and directories. It also manages the mapping of blocks to DataNodes.

DataNode

DataNodes represent the slave in the architecture that manages data and the storage attached to it. A typical HDFS cluster can have thousands of DataNodes and tens of thousands of HDFS clients per cluster, since each DataNode may execute multiple application tasks simultaneously. The DataNodes are responsible for managing read and write requests from the filesystem's clients and block maintenance and replication as directed by the NameNode. The block management in HDFS is different from a normal filesystem. The size of the data file equals the actual length of the block. This means if a block is half full it needs only half of the space of the full block on the local drive, thereby optimizing storage space for compactness, and there is no extraspace consumed on the block unlike a regular filesystem.

A filesystem-based architecture needs to manage consistency, recoverability, and concurrency for reliable operations. HDFS manages these requirements by creating image, journal, and checkpoint files.

Image

An image represents the metadata of the namespace (inodesand lists of blocks). On startup, the NameNode pins the entire namespace image in memory. The in-memory persistence enables the NameNode to service multiple client requests concurrently.

Journal

The Journal represents the modification log of the image in the local host's native filesystem. During normal operations, each client transaction is recorded in the journal, and the journal file is flushed and synced before the acknowledgment is sent to the client. The NameNode upon startup or from a recovery can replay this journal.

Checkpoint

To enable recovery, the persistent record of the image is also stored in the local host's native files system and is called a checkpoint. Once the system starts-up, the NameNode never modifies or updates the checkpoint file. A new checkpoint file can be created during the next startup, on a restart or on demand when requested by the administrator or by the CheckpointNode (described later in this chapter).

HDFS startup

Since the image is an in-memory persistence, during initial startup everytime, the NameNode initializes a namespace image from the checkpoint file and replays changes from the journal. Once the startup sequence completes the process, a new checkpoint and an empty journal are written back to the storage directories and the NameNode starts serving client requests. For improved redundancy and reliability, copies of checkpoint and journal can be made at other servers.

Block allocation and storage

Data organization in the HDFS is managed similar to GFS. The namespace is represented by inodes, which represent files. Directories and records attributes like permissions, modification, and access times, namespace and disk space quotas. The files are split into use-defined block sizes (default is 128 MB) and stored into a DataNode and two replicas at a minimum to ensure availability and redundancy, though the user can configure more replicas. Typically the storage location of block replicas may change over time and hence are not part of the persistent checkpoint.

HDFS client

A thin layer of interface that is used by programs to access data stored within HDFS, is called the Client. The client first contacts the NameNode to get the locations of data blocks that comprise the file. Once the block data is returned to the client, subsequently the client reads block contents from the DataNode closest to it.

When writing data, the client first requests the NameNode to provide the DataNodes where the data can be written. The NameNode returns the block to write the data. When the first block is filled, additional blocks are provide by the NameNode in a pipeline. A block for each request might not be on the same DataNode.

One of the biggest design differentiators of HDFS is the API that exposes the locations of a file blocks. This allows applications like MapReduce to schedule a task to where the data is located, thus improving the IO performance. The API also includes functionality to set the replication factor for each file. To maintain file and block integrity, once a block is assigned to a DataNode, two files are created to represent each replica in the local host's native filesystem. The first file contains the data itself and the second file is block's metadata including checksums for each data block and generation stamp.

Replication and recovery

In the original design of HDFS there was a single NameNode for each cluster, which became the single point of failure. This has been addressed in the recent releases of HDFS where NameNode replication is now a standard feature like DataNode replication.

NameNode and DataNode—communication and management

The communication and management between a NameNode and DataNodes are managed through a series of handshakes and system ID's. Upon initial creation and formatting, a namespace ID is assigned to the filesystem on the NameNode. This ID is persistently stored on all the nodes across the cluster. DataNodes similarly are assigned a unique storage_idon the initial creation and registration with a NameNode. This storage_id never changes and will be persistent event if the DataNode is started on a different IP Address or Port.

During startup process, the NameNode completes its namespace refresh and is ready to establish the communication with the DataNode. To ensure that each DataNode that connects to the NameNode is the correct DataNode, there is a series of verification steps:

- The DataNode identifies itself to the NamNode with a handshake and verifies its namespace ID and software version.
- If either does not match with the NameNode, the DataNode automatically shuts down.
- The signature verification process prevents incorrect nodes from joining the cluster and automatically preserves the integrity of the filesystem.
- The signature verification process also is an assurance check for consistency of software versions between the NameNode and DataNode since incompatible version can cause data corruption or loss.
- Post the handshake and validation on the NameNode, a DataNode sends a block report. A block report contains the block id, the length for each block replica and the generation stamp.
- The first block report is sent immediately upon the DataNode registration.
- Subsequently hourly updates of the block report is sent to the NameNode, which provides the view of where block replicas are located on the cluster.
- When a new DataNode is added and initialized, since it does not have a namespace ID is permitted to join the cluster and receive the cluster's namespace ID.

Heartbeats

The connectivity between the NameNode and DataNode are managed by the persistent heartbeats that are sent by the DataNode every 3 seconds. The heartbeat provides the

NameNode confirmation about the availability of the blocks and the replicas of the DataNode. Additionally,heartbeats also carry information about total storage capacity, storage in use, and the number of data transfers currently in progress. These statistics are by the NameNode for managing space allocation and load balancing.

During normal operations, if the NameNode does not receive a heartbeat from a DataNode in 10 minutes, the NameNodeconsiders theDataNode to be out of service and the block replicas hosted to be unavailable. The NameNode schedules creation of new replicas of those blocks on other DataNodes.

The heartbeats carry round-trip communications and instructions from the NameNode, these include commands to

- Replicate blocks to other nodes
- Remove local block replicas
- Reregister the node
- Shut down the node
- Send an immediate block report

Frequent heartbeats and replies are extremely important for maintaining the overall system integrity even on big clusters. Typically a NameNode can process thousands of heartbeats per second without affecting other operations.

CheckPointNode and BackupNode

There are two roles that a NameNode can be designated to perform apart from servicing client requests and managing DataNodes. These roles are specified during startup and can be the CheckPointNode or the BackupNode.

CheckPointNode

The CheckpointNode serves as a journal capture architecture to create a recovery mechanism for the NameNode. The checkpointnode combines the existing checkpoint and journal to create a new checkpoint and an empty journal in specific intervals. It returns the new checkpoint to the NameNode. The CheckpointNode will runs on a different host from the NameNode since it has the same memory requirements as the NameNode.

By creating a checkpoint the NameNode can truncate the tail of the current journal. Since HDFS clusters run for prolonged periods of time without restarts, resulting in very large journal growth, increasing the probability of loss or corruption. This mechanism provides a protection mechanism.

BackupNode

The BackupNode can be considered as a read-only NameNode. It contains all filesystem's metadata information except for block locations. It accepts a stream of namespace transactions from the active NameNode and saves them to its own storage directories, and applies these transactions to its own namespace image in its memory. If the NameNode fails, the BackupNode's image in memory and the checkpoint on disk is a record of the latest namespace state and can be used to create a checkpoint for recovery. Creating a checkpoint from a BackupNode is very efficient as it processes the entire image in its own disk and memory.

A BackupNode can perform all operations of the regular NameNode that does not involve modification of the namespaceor management of block locations. This feature provides the administrators the option of running a NameNode without persistent storage, delegating responsibility for the namespace state persisting to the BackupNode. This is not a normal practice, but can be used in certain situations.

Filesystem snapshots

Like any filesystem, there are periodic upgrades and patches that might need to be applied to the HDFS. The possibility of corrupting the system due to software bugs or human mistakes always exists. In order to avoid system corruption or shutdown, we can create snapshots in HDFS. The snapshot mechanism lets administrators save the current state of the filesystem to create a rollback in case of failure.

Load balancing, disk management, block allocation, and advanced file management are topics handled by HDFS design. For further details on these areas, refer to the HDFS architecture guide on Apache HDFS project page.

Based on the brief architecture discussion of HDFS, we can see how Hadoop achieves unlimited scalability and manages redundancy while keeping the basic data management functions managed through a series of API calls.

MapReduce

We discussed earlier in the chapter on the pure MapReduce implementations on GFS, in the big data technology deployment, Hadoop is the most popular and deployed platform for the MapReduce framework. There are three key reasons for this:

- Extreme parallelism available in Hadoop
- Extreme scalability programmable with MapReduce
- The HDFS architecture

To run a query or any procedural language like Java or C++ in Hadoop, we need to execute the program with a MapReduce API component. Let us revisit the MapReduce components in the Hadoop architecture to understand the overall design approach needed for such a deployment.

YARN—yet another resource negotiator

The advancements of Hadoop were having an issue in 2011, the focus of the issue was highlighted by Eric Baldeschwieler the then CEO of Hortonworks when MapReduce distinctly showcased two big areas of weakness one being scalability and second the utilization of resources. The goal of the new framework which was titled Yet Another Resource Negotiator (YARN) was to introduce the operating system for Hadoop. An operating system in Hadoop ensures scalability, performance, and resource utilization which has resulted in an architecture for Internet of Things to be implemented. The most important concept of YARN is the ability to implement a data processing paradigm called as lazy evaluation and extremely late binding (we will discuss this in all the following chapters), and this feature is the future of data processing and management. The ideation of a data warehouse will be very much possible with an operating system model where we can go from raw and operational data to data lakes and data hubs.

YARN addresses the key issues of Hadoop 1.0, and these include the following:

- The JobTracker is a major component in data processing as it manages key tasks of resource marshaling and job execution at individual task levels. This interface has deficiencies in
  - Memory consumption
  - Threading-model
  - Scalability
  - Reliability
  - Performance

These issues have been addressed by individual situations and several tweaks in design are done to circumvent the shortcomings. The problem manifests in large clusters where it becomes difficult to manage the issue (Fig. 2.8).

- Overall issues have been observed in large clustered environments in the areas of
  - Reliability
  - Availability
  - Scalability—Clusters of 10,000 nodes or/and 200,000 cores
  - Evolution—Ability for customers to control upgrades to the grid software stack
  - Predictable Latency—A major customer concern
  - Cluster utilization
  - Support for alternate programming paradigms to MapReduce

The two major functionalities of the JobTracker are resource management and job scheduling/monitoring. The load that is processed by JobTracker runs into problems due to competing demand for resources and execution cycles arising from the single point of control in the design. The fundamental idea of YARN is to split up the two major functionalities of the JobTracker into separate processes. In the new release architecture,

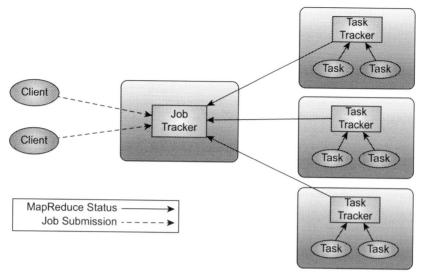

FIGURE 2.8 MapReduce classic — JobTracker architecture. *Image source—Apache Foundation.*

there are two modules a global ResourceManager (RM) and per-application ApplicationMaster (AM) (Fig. 2.9).

The primary components and their focus areas are as follows:

- ResourceManager (RM)
  - Has two main components
    - Scheduler
      - The Scheduler is responsible for allocating resources to the various running applications and manages the constraints of capacities, availability, and resource queues
      - The Scheduler will be responsible for purely schedule management and will be working on scheduling based on resource containers, which specify memory, disk, and CPU
      - Scheduler will not assume restarting of failed tasks either due to application failure or hardware failures
    - Applications Manager (AM)
      - Responsible for accepting jobsubmissions
      - Negotiates the first container for executing the application-specific ApplicationMaster
      - Provides the service for restarting the ApplicationMaster container on failure
      - Applications Manager has three sub components:
        - Scheduler negotiator—Component responsible for negotiating the resources for the AM with the Scheduler

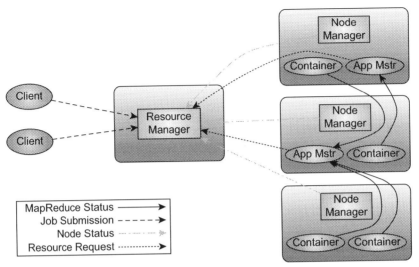

FIGURE 2.9 YARN architecture. *Image source—Apache Foundation.*

  ⁕ AMContainer Manager—Component responsible for starting and stop-
    ping the container of the AM by talking to the appropriate
    NodeManager
  ⁕ AM Monitor—Component responsible for managing the aliveness of
    the AM and responsible for restarting the AM if necessary

   The resource manager stores snapshots of its state in the Zookeeper. In case of failure,
a very transparent restart is feasible and ensures availability.

- Node Manager
  ⁕ Is a per-machine agent and is responsible for launching containers for applica-
    tions once the scheduler allocates them to the application.
  ⁕ Container resource monitoring for ensuring that the allocated containers do not
    exceed their allocated resource slices on the machine.
  ⁕ Setting up the environment of the container for the task execution including bi-
    naries, libraries, and jars
  ⁕ Manages local storage on the node. Applications can continue to use the local
    storage even when they do not have an active allocation on the node, thus
    providing scalability and availability
- ApplicationMaster
  ⁕ Perapplication
  ⁕ Negotiates resources with ResourceManager
  ⁕ Manages application scheduling and task execution with Node Managers
  ⁕ Recovers the application on its own failure. Will either recover the application
    from saved persistent state or just run the application from the very beginning,
    depending on recovery success

YARN scalability

The resource model for YARN is memory driven. Every node in the system is modeled to be consisting of multiple containers of minimum size of memory. The ApplicationMaster can request multiple of the minimum memory size as needed.

What this means to any application is the memory slots required to run a job can be accessed from any node, depending on the availability of memory. This provides simple chunkable scalability especially in a cluster configuration. In classic Hadoop MapReduce the cluster is not artificially segregated into map and reduce slots and the application jobs are bottlenecked on reduce slots limiting scalability in job execution in the dataflow (Fig. 2.10).

YARN execution flow

Comparison between MapReduce v1 and v2

Presented here is a simple comparison between the two releases of MapReduce

classic MapReduce	YARN
- Job request submitted to JobTracker	- Application executed by YARN
- Jobtracker manages the execution with tasks	- Resources negotiated and allocated prior to job execution
- Resources are allocated on availability basis, some jobs get more and others less	- Map based resource request setup for the entire job
- Resource allocation across a cluster	- Resource monitor tracks usage and requests additional resource as needed from across a cluster in a clustered setup
- Multiple single points of failure	- Job completion and cleanup tasks are executed

SQL/MapReduce

Business intelligence has been one of the most successful applications in the last decade, but severe performance limitations have been a bottleneck especially with detailed data analysis. The problem becomes compounded with analytics and the need for 360 degrees perspective on customer and product with ad-hoc analysis demands from users. The powerful combination of SQL when extended to MapReduce will enable users to explore larger volumes of raw data through normal SQL functions and regular BI tools. This is the fundamental concept behind SQL/MapReduce. There are a few popular implementations of SQL/MapReduce including Hive, AsterData, Greenplum, and HadoopDB.

Fig. 2.5 shows a conceptual architecture of an SQL/MapReduce implementation. There are a few important components to understand:

- Translator—this is a custom layer provided by the solution. It can simply be a library of functions to extend in the current database environment

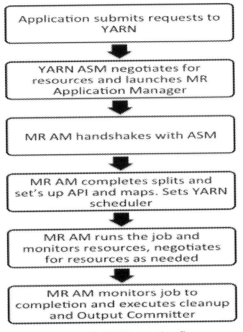

FIGURE 2.10 YARN execution flow.

- SQL/MapReduce interface—this is the layer that will create and distribute the jobs at the lowest MapReduce execution layer for
- SQL/MapReduce Libraries—catalog of library functions

    The overall benefits of combining SQL/MapReduce include the following (Fig. 2.11):

- Use of SQL for powerful postresult analytics and MapReduce to perform large-scale data processing on unstructured and semistructured data
- Effectively use the sharding capabilities of MapReduce to scale up and scale out the data irrespective of volume or variety
- Provide the business user all the data with the same interface tool that runs on SQL

    The downside of the technology in evolution includes the following:

- Heavy dependency on custom libraries
- Current support on certain analytic functions

    The next generation of SQL/Mapreduce interfaces and libraries will solve a number of evolutionary challenges.

    The combination of HDFS and MapReduce creates an extreme architecture. What is important to note here is as follows:

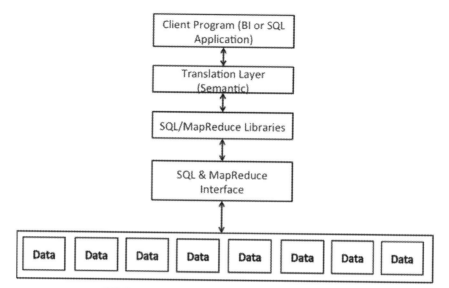

FIGURE 2.11 Conceptual SQL/MapReduce architecture.

- Files once processed cannot be processed from a mid-point. If a new version of the data is sent by files, the entire file has to be processed
- MapReduce on large clusters can be difficult to manage
- The entire platform by design is oriented to handle extremely large files and hence is not suited for transaction processing
- When the files are broken for processing, the consistency of the files completing processing on all nodes in a cluster is a soft state model of eventual consistency

Zookeeper

Developing large-scale applications on Hadoop or any distributed platform mandates that a resource and application coordinator be available to coordinate the tasks between nodes. In a controlled environment like the RDBMS or SOA programming, the tasks are generated in a controlled manner and the coordination simply needs to ensure successful network management without data loss and the health check on the nodes in a distributed system. In the case of Hadoop, the minimum volumes of data starts with multi-terabytes and the data is distributed across files on multiple nodes. Keeping users queries and associated tasks mandates a coordinator that is as flexible and scalable as the platform itself.

ZooKeeper is an open source, in-memory, distributed NoSQL database that is used for coordination services for managing distributed applications. It consists of a simple set of functions that can be used to build services for synchronization, configuration maintenance, groups, and naming. Zookeeper has a filesystem structure that mirrors classic filesystem tree architectures; it is natively developed and deployed in Java and has bindings for Java and C.

Because of its NoSQL origins, the architecture of Zookeeper is based on a hierarchical namespace structure (called zNodes). It uses the zNode architecture to allow distributed processes to coordinate and have interprocess communication mechanisms, in a low latency, high throughput, and ordered access environment. Sophisticated synchronization is implemented at the client by using a feature called ordering based on vector clocks and timestamps. ZooKeeper is primarily designed to store coordination data: status information, configuration, location information, etc and not large data volumes. By default, ZooKeeper has a built-in check for a maximum size of 1 MB.

Zookeeper features

Data Model—A Zookeeper model is simple to understand

- A parent namespace called zNode
  - Contains children and further levels of zNodes
  - Can have data associated with any level
  - Has Access Control Lists
  - A session based node called Ephemeral node can be created and destroyed by a session
  - zNodes can be sequential, this ordering helps manage complex interconnected tasks
  - zNodes can have watches to provide for callbacks. This helps when data is distributed and tasks are distributed in a large cluster
- API—Zookeeper has a simple set of API commands to work with
  - string create(path, data, acl, flags)
  - delete(path, expected_version)
  - statset_data(path, data, expected_version)
  - (data, stat) get_data(path, watch)
  - stat exists(path, watch)
  - stringget_children(path, watch)
- Zookeeper Consistency—Zookeeper offers a data consistency model to provide guarantees with its services
  - Sequential Consistency—Updates from a client are applied in the order sent
  - Atomicity—Updates either succeed or fail
  - Single System Image—Unique view, regardless of the server
  - Durability—Updates once succeeded will not be undone
  - Timeliness—Lag is bounded, read operations can lag leaders
- Used for
  - Configuration service —Get latest configuration and get notified changes occur
  - Lock service—Provide mutual exclusion

* Leader election—There can be only one
* Group membership—Dynamically determine members of a group
* Queue Producer/Consumer paradigm

In Hadoop ecosystem, Zookeeper is implemented as a service to coordinate tasks. Fig. 2.12 shows the implementation model of Zookeeper.

ZooKeeper as a service can be run in two modes:

* Standalone mode—there is a single ZooKeeper server and this configuration is useful for development or testing but provides no guarantees of highavailability or resilience.
* Replicated mode—this is the mode of deployment in production, on a cluster of machines called an ensemble. ZooKeeper achieves high availability through replication, and can provide a service as long as a majority of the machines in the ensemble are up and running. For example as seen in Fig. 2.12, in a five-node ensemble, any two machines can fail and the service will still work because a majority of three remains (a quorum), whereas in a six node ensemble, a failure of three means loss of majority, and shutdown of service. It is usual to have an odd number of machines in an ensemble to avoid such situations.

Zookeeper has one task or goal, to endure that all the zNode changes across the system are updated to the leader and followers. When a failure occurs in a minority of machines, the replicas need to bring up the machines to a catch-up from the lag. To implement the management of the ensemble, ZooKeeper uses a protocol called Zab that runs in two steps and can be repetitive.

1. Leader election—The machines in an ensemble go through a process of electing a distinguished member, called the leader. Clients communicate with one server in a session and work on a read or write operation. As seen here, writes will be only accomplished through the leader, which is then broadcast to the followers as an

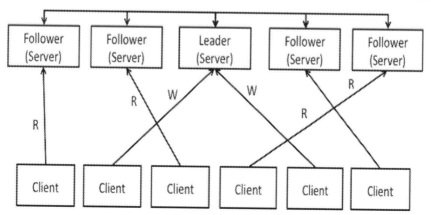

FIGURE 2.12 Zookeeper ensemble.

update. Reads can be from the leader or followers and happen in memory. Followers sometimes lag in read operations and eventually become consistent. This phase is finished once a majority (or quorum) of followers have synchronized their state with the leader.

Zab implements the following optimizations to circumvent the bottleneck of a leader:

⚬ Clients can connect to any server, and servers have to serve read operations locally and maintain information about the session of a client. This extra load of a follower process (a process that is not a leader) makes the load more evenly distributed

⚬ The number of servers involved is small. This means that the network communication overhead does not become the bottleneck that can affect fixed sequencer protocols

2. Atomic broadcast—Written requests and updates are committed in a two-phase approach in Zab. To maintain consistency across the ensemble, write requests are always communicated to the leader. The leader broadcasts the update to all its followers. When a quorum of its followers (in Fig. 2.12 we need three followers) have persisted the change (phase 1 of a two-phase commit), the leader commits the update (phase 2 of the commit), and the requestor gets a response saying the update succeeded. The protocol for achieving consensus is designed to be atomic, so a change either succeeds or fails completely.

Locks and processing

One of the biggest issues in distributed data processing is lock management, when one session has an exclusive lock on a server. Zookeeper manages this process by creating a list of child nodes and lock nodes and the associated queues of waiting processes for lock release. The lock node is allocated to the next process that is waiting based on the order received.

Lock management is done through a set of watches. If you become overzealous and set a large number of locks it will become a nightmare and creates a herd effect on the Zookeeper service. Typically a watch is set on the preceding process that is currently holding a lock.

Failure and recovery

A common issue in coordinating a large number of processes is connection loss. When failover process needs information on the children affected by the connection loss to complete the failover. To manage this the client session id information is associated with child zNodes and locks, which will enable the failover client to synchronize.

ZooKeeper is a highly available system, and it is critical that it can perform its functions in a timely manner. It is recommended to run ZooKeeper on dedicated machines. Running it in a shared services environment will adversely impact performance.

In Hadoop deployment, Zookeeper serves as the coordinator for managing all the key activities:

- **Manage configuration across nodes**—ZooKeeper helps you quickly push configuration changes across dozens or hundreds of nodes
- **Implement reliable messaging**—A guaranteed messaging architecture to deliver messages can be implemented with Zookeeper.
- **Implement redundant services**—Managing a large number of nodes with a Zab approach will provide a scalable redundancy management solution.
- **Synchronize process execution**—With ZooKeeper, multiple nodes can coordinate the start and end of a process or calculation. This approach can ensure consistency of completion of operations.

Pig

Analyzing large data sets introduces data flow complexities that become harder to implement in a MapReduce program as data volumes and processing complexities increase. A high-level language that is more user friendly and is SQL-like in terms of expressing data flows and has the flexibility to manage multi-step data transformations, handle joins with simplicity, and easy program flow was needed as an abstraction layer over MapReduce.

Apache Pig is a platform that has been designed and developed for analyzing large data sets. Pig consists of a high-level language for expressing data analysis programs and comes with infrastructure for evaluating these programs. At the time of writing this book (2012), Pig's current infrastructure consists of a compiler that produces sequences of MapReduce programs. Pig's language architecture is a textual language platform called Pig Latin, whose design goals were based on the requirement to handle large data processing with minimal complexity and include the following:

- **Programming Flexibility**—ability to break down complex tasks comprised of multiple steps and interprocess-related data transformations should be encoded as data flow sequences that are easy to design, develop, and maintain.
- **Automatic Optimization**—Tasks are encoded to let the system optimize their execution automatically. This allows the user with greater focus on program development allowing the user to focus on semantics rather than efficiency.
- **Extensibility**—Users can develop UDFs for more complex processing requirements

Programming with Pig Latin

Pig is primarily a scripting language for exploring large datasets. It is developed to process multiple terabytes of data in half-dozen lines of Pig Latin code. Pig provides several commands to the developer for introspectingthe data structures in the program, as it is written.

Pig Latin can be executed as statements in either In Local or MapReduce mode either interactively or as batch programs.

- In local mode, Pig runs in a single JVM and accesses the local filesystem. This mode is suitable only for small datasets and can be run on minimal infrastructure.
- In MapReduce mode Pig translates programs (queries and statements) into MapReduce jobs and runs them on a Hadoop cluster. Production environments for running Pig are deployed in this mode

Pig data types

Pig language supports the following data types:

- Scalar types: int, long, double, chararray, bytearray
- Complex types:
- map: associative array
- tuple: ordered list of data, elements may be of any scalar or complex type
- bag: unordered collection of tuples

Running Pig programs

Pig programs can be run in three modes, all of which work in both local and
MapReduce mode (for more details see Apache Pig Wiki Page):
Scripting Driven—A Pig program can be run as a script file, processed from command line
Grunt Shell—An interactive shell for running Pig commands
Embedded—You can run Pig programs from Java, using JDBC drivers like a traditional SQL programs from Java.

Pig program flow

Pig program control has many built-in commands and syntax. We will take a look at the core execution model. Every Pig module has the LOAD, DUMP, and STORE statement.

- A LOAD statement reads data from the filesystem.
- A series of "transformation" statements process the data
- An STORE statement writes output to the filesystem
- A DUMP statement displays output to the screen

Common Pig command

LOAD—Read data from filesystem
STORE—Write data to filesystem

FOREACH—Apply expression to each record and output one or more records
FILTER—Apply predicate and remove records that do not return true
GROUP/COGROUP—Collect records with the same key from one or more inputs
JOIN—Join two or more inputs based on a key
ORDER—Sort records based on a key
DISTINCT—Remove duplicate records
UNION—Merge two data sets
SPLIT—Split data into two or more sets, based on filter conditions
STREAM—Send all records through a user provided binary
DUMP—Write output to stdout
LIMIT—Limit the number of records

During program execution, Pig first validates the syntax and semantics of statements and continues to process them, when it encounters a DUMP or STORE it completes the execution of the statement. For example, a Pig job to process compliance logs and extract words and phrases will look like

```
A = load "compliance_log"
B = foreach A generate
flatten(TOKENIZE((chararray)$0)) as word;
C = filter B by word matches '\\w+';
D = group C by word;
E = foreach D generate COUNT(C), group;
store E into "ompliance_log_freq";
```

Now let us say that we want to analyze how many of these words are in FDA mandates

```
A = load "FDA_Data";
B = foreach A generate
flatten(TOKENIZE((chararray)$0)) as word;
C = filter B by word matches "\\w+";
D = group C by word;
E = foreach D generate COUNT(C), group;
store E into "FDA_Data_freq";
```

We can then join these two outputs to create a result set:

```
compliance = LOAD "compliance_log_freq" AS (freq, word)
FDA = LOAD "FDA_Data_freq" AS (freq, word)
inboth = JOIN compliance BY word, FDA BY word
STORE inboth INTO "output";
```

In this example the FDA data is highly semistructured and compliance logs are generated by multiple applications. Processing large data with simple lines of code is what Pig brings to MapReduce and Hadoop data processing.

Pig can be used more in data collection and preprocessing environments and in streaming data processing environments. It is very useful in data discovery exercises.

HBASE

HBASE is an open source, nonrelational database modeled on Google's Big Table architecture, completely developed in Java. It runs on Hadoop and HDFS providing real time read/write access to large data sets on Hadoop platform. HBASE is not a database in a purist definition of the database. It provides unlimited scalability and performance for RDBMS like capabilities while not being ACID compliant. HBASE has been classified as a NOSQL database as it is modeled after Google Bigtable.

HBASE architecture

Data is organized in HBASE as rows and columns, and tables, very similar to a database, here is where the similarity ends. Let us look at the data model of HBASE and then understand the implementation architecture.

Data Model—A data model of HBASE consists of Tables, Column Groups, and Rows.

- Tables
  - ⁂ Tables are made of rows and columns
  - ⁂ Table cells—are the intersection of row and column coordinates. Each cell is versioned by default with a Timestamp. The contents of a cell are treated as uninterpretedarray of bytes
  - ⁂ A Table row has a sortable rowkey and an arbitrary number of columns
- Rows
  - ⁂ Table rowkeys are also byte arrays. In this configuration anything can serve as the rowkey opposed to strongly typed datatypes in the traditional database
  - ⁂ Table rows are sorted byte-ordered by rowkey, the table's primary key, all table accesses are via the table primary key.
  - ⁂ Columns are grouped as families and a row can have as many columns as loaded
- Columns and Column Groups (families)
  - ⁂ In HBASE row columns are grouped into column families
  - ⁂ All column family members will mandatorily have a common prefix, for example, the columns person:name and person:comments are both members of the person column family, whereas, email:identifier belongs to the email family
  - ⁂ A table's column families must be specified up front as part of the table schema definition,
  - ⁂ New column family members can be added on demand.

The flexibility of this type of a data model organization allows HBASE to store data as column oriented grouped by column families by design. The columns can be expanded based on the data loaded as long as it belongs to the row that it is loaded to and has predefined column groups in the data model.

As seen in Fig. 2.13 there are multiple records for one rowkey and one row for the other. The flexibility to store this data in column groups allows us to store more data and query more data in the same query cycle. These are the powerful data model structures that are implemented in a larger architecture within Hadoop.

HBASE architecture implementation

There are three main components that together form the HBASE architecture from an implementation model perspective (Fig. 2.14)

- The HBaseMaster—is the key controller of operations in HBASE. The main functions of the master include the following:
  * Responsible for monitoring region servers (one or more clusters)
  * Load balancing for regions
  * Redirect client to correct region servers

To manage redundancy the master can be replicated. Like the master in MapReduce, the master in HBASE stores no data and only has metadata about the region servers.

- The HRegionServer—is the slave node in the HBASE architecture. Its primary functions include the following:
  * Storing the data and its metadata
  * Serving requests(Write/Read/Scan) of client
  * Send heartbeat to master
  * Manage splits and synchronize with master on the split and data allocation

Row key	TS	Column "recipe:"	
"www.foodie.com"	t10	"recipe: foodie.com"	"FOODIE"
"www.foodtv.com"	t9	"recipe: foodtv.com"	"FOODTV.COM"
	t8	"recipe: food.spicy.in"	"FOODTV.COM"

FIGURE 2.13 HBASE data model example.

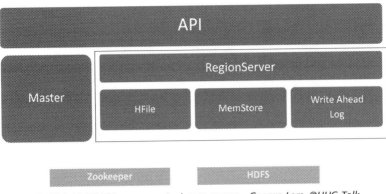

FIGURE 2.14 HBASE components. *Image source—George Lars, @HUG Talk.*

- The HBase client is a program API that can be executed from any language like Java or C++ to access HBASE
- Zookeeper—HBASE uses Zookeeper to coordinate all the activities between master and region servers

How does HBASE internally manage all the communication between Zookeeper, master servers, and region servers? HBASE maintains two special catalog tables named ROOT and META. It maintains the current list, state, and location of all regions afloat on the cluster in these two catalogs. ROOT table contains the list of META table regions, and META table contains the list of all userspace regions. Entries in ROOT and META tables are keyed by region names, where a region name is made of the table name the region belongs to, the region's start row, its time of creation, and a hash key value. Rowkeys are sorted by default and finding the region that hosts a particular row is a matter of a lookup to find the first entry where the key is greater than or equal to that of the requested rowkey. AS regions are split or deleted or disabled, the ROOT and META tables are constantly refreshed and thus the changes are immediately reflected to user requests.

Clients connect to the ZooKeeper and get the access information to the ROOT. The ROOT provides information about the META, which points to the region whose scope covers that of the requested row. The client then gets all the data about the region, user space, the column family, and the location details by doing a lookup on the META table. Post the initial interaction with the master, the client directly starts working with the hosting region server.

HBASE Clients cache all the information they gather traversing ROOT and META, by caching locations as well as the userspace, the region start and stop rows. The cached data provides all the details about the regions and the data available there, avoiding round trips to read the META table. In a normal mode of operation, clients continue to use the cached entries as they perform tasks, until there is a failure or abort. When a failure happens, it is normally due to the movement of the region itself causing the cache to become stale. When this happens, the client first queries the META and if the META table itself has moved to another location, the client traverses back to the ROOT table to get further information.

When clients write data to HBASE tables, this data is first processed inmemory. When the memory becomes full, the data is flushed to a log file. The file is available on HDFS for use by HBASE in crash recovery situations.

HBASE is a powerful column-oriented datastore, which is truly a sparse, distributed, persistent multidimensional sorted map. It is the database of choice for all Hadoop deployments as it can hold the key–value outputs from MapReduce and other sources in a scalable and flexible architecture.

Hive

The scalability of Hadoop and HDFS is unparalleled based on the underlying architecture and the design of the platform. While HBASE provides some pseudo database features, business users working on Hadoop did not adopt the platform due to lack of SQL support or SQL-like features on Hadoop. While we understand that Hadoop cannot answer low latency queries and deep analytical functions like the database, it has large data sets that cannot be processed by the database infrastructure and needs to be harnessed with some SQL-like language or infrastructure that can run MapReduce in the internals. This is where HIVE comes into play.

Hive is an opensource data warehousing solution that has been built on top of Hadoop. The fundamental goals of designing Hive are as follows:

• To build a system for managing and querying data using structured techniques on Hadoop
• Use native MapReduce for execution at HDFS and HADOOP layers
• Use HDFS for storage of Hive data
• Store key metadata in an RDBMS
• Extend SQL Interfaces—Familiar data warehousing tool in use at enterprises
• High Extensibility—User-defined types, user-defined functions, formats, scripts
• Leverage extreme scalability and performance of Hadoop
• Interoperability with other platforms

Hive supports queries expressed in an SQL-like declarative language—HiveQL, which are compiled into MapReduce jobs executed on Hadoop. Hive also includes a system catalog, metastore, which contains schemas and statistics and is used in data exploration and query optimization.

Hive was originally conceived and developed at Facebook when the data scalability needs of Facebook outpaced and outgrew any traditional solution. Over the last few years, Hive has been released as an open source platform on the Apache Hadoop project. Let us take a quick look at Hive architecture (Fig. 2.15).

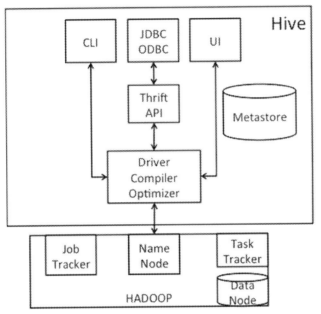

FIGURE 2.15 Hive architecture.

Hive architecture

Hive system architecture is seen in Fig. 2.12. The main architecture components that form the building blocks are as follows:

- **Metastore**—Stores the system catalog and metadata about tables, columns, and partitions
- **Driver**—Maintains session details, process handles, statistics, and manages the life-cycle of a HiveQL statement as it moves through Hive.
- **Query Compiler**—Compiles HiveQL into map/reduce tasks (follows a DAG model)
- **Execution Engine**—Processes and executes the tasks produced by the compiler in a dependency order. The execution engine manages all the interactions between the compiler and Hadoop
- **Thrift Server**—Provides a thrift interface, a JDBC/ODBC server, and a rich API to integrate Hive with other applications.
- **CLI and Web UI**—Are two client interfaces. Command line interface allows command line execution and the web user interface is a management console.
- **Interfaces**—Extensibility interfaces include the SerDe (implemented as Lazy SerDe in Hive) and ObjectInspector, UDF(user-defined function) and UDAF(user-defined aggregate function) interfaces that enable users to define their own custom functions.

The next section describes in detail the functionality of the components broken down into infrastructure and execution.

Infrastructure

- Metastore—The metastore is the system catalog which contains metadata about the tables stored in Hive. Metadata is specified during table creation and reused everytime the table is used or specified in HiveQL. Metastore can be compared to a system catalog in a traditional database speak. The metastore contains the following objects:
- Database—is the default namespace for tables. Users can create a database and name it. The database "default" is used for tables when no user supplied database name.
- Table—A Hive table is made up of the data being stored in it and the associated metadata metastore.
  - In the physical implementation the data typically resides in HDFS, although it may be in any Hadoop filesystem, including the local filesystem.
  - Metadata for table typically contains the list of columns and their data types, owner, user-supplied keys, storage, and SerDe information.
  - Storage information includes location of the table's data in the filesystem, data formats, and bucketing information.
  - SerDe metadata includes the implementation class of serializer and deserializer methods and any supporting information required by that implementation.
  - All this information can be specified during the initial creation of table.
- Partition—In order to gain further performance and scalability, Hive organizes tables into partitions.
  - A partition contains parts of the data, based on the value of a partition column, for example date or LatLong.
  - Tables or partitions can be further subdivided into buckets. A bucket is akin to a subpartition. An example is to bucket a partition of customers by customer_id.
  - Each partition can have its own columns and SerDe and storage information.

Execution—how does Hive process queries?

A HiveQL statement is submitted via the CLI, the web UI, or an external client using the Thrift, ODBC, or JDBC API. The driver first passes the query to the compiler where it goes through parse, type check, and semantic analysis using the metadata stored in the metastore. The compiler generates a logical plan that is then optimized through a simple rule–based optimizer. Finally an optimized plan in the form of a DAG of mapreduce tasks and HDFS tasks is generated. The execution engine then executes these tasks in the order of their dependencies, using Hadoop.

We can further analyze this workflow of processing as follows:

- Hive client triggers a query
- Compiler receives the query and connects to metastore
- Compiler receives the query and initiates the first phase of compilation
  - Parser—Converts the query into parse tree representation. Hive uses ANTLR to generate the abstract syntax tree (AST)
  - Semantic Analyzer—In this stage the compiler builds a logical plan based on the information that is provided by the metastore on the input and output tables. Additionally the complier also checks type compatibilities in expressions and flags compile time semantic errors at this stage. The best step is the transformation of an AST to intermediate representation that is called the query block (QB) tree. Nested queries are converted into parent—child relationships in a QB tree during this stage
  - Logical Plan Generator—In this stage the compiler writes the logical plan from the semantic analyzer into a logical tree of operations
  - Optimization—This is the most involved phase of the complier as the entire series of DAG optimizations are implemented in this phase. There are several customizations than can be done to the complier if desired. The primary operations done at this stage are as follows:
    - Logical optimization—Perform multiple passes over logical plan and rewrites in several ways
    - Column pruning—This optimization step ensures that only the columns that are needed in the query processing are actually projected out of the row
    - Predicate pushdown—Predicates are pushed down to the scan if possible so that rows can be filtered early in the processing
    - Partition pruning—Predicates on partitioned columns are used to prune out files of partitions that do not satisfy the predicate
    - Join optimization
    - Grouping and regrouping
    - Repartitioning
    - Physical plan generator converts logical plan into physical.
    - Physical plan generation creates the final DAG workflow of MapReduce
  - Execution engine gets the compiler outputs to execute on the Hadoop platform.
    - All the tasks are executed in the order of their dependencies. Each task is only executed if all of its prerequisites have been executed.
    - A map/reduce task first serializes its part of the plan into a plan.xml file.
    - This file is then added to the job cache for the task and instances of ExecMapper and ExecReducers are spawned using Hadoop.
    - Each of these classes deserializes the plan.xml and executes the relevant part of the task.

- The final results are stored in a temporary location and at the completion of the entire query, the results are moved to the table if inserts or partitions, or returned to the calling program at a temporary location

The comparison between how Hive executes versus a traditional RDBMS shows that due to the schema on read design, the data placement, partitioning, joining, and storage can be decided at the execution time rather than planning cycles.

Hive data types

Hive supports the following data types—tinyint, int, smallint, bigint, float, boolean, string, and double. Special data types include Array, Map(key–value pair), and Struct (collection of names fields).

Hive query language (HiveQL)

The Hive query language (HiveQL) is an evolving system that supports a lot of SQL functionality on Hadoop, abstracting the MapReduce complexity to the end users.

Traditional SQL features like select, create table, insert, "from clause" subqueries, various types of joins—inner, left outer, right outer and outer joins, "group by"and aggregations, union all, create table as select, and many useful functions.

Hive examples

Count Rows in a table –
    SELECT COUNT(1) FROM table2;
    SELECT COUNT(*) FROM table2;
    Order By - colOrder: (ASC | DESC)
    orderBy: ORDER BY colNamecolOrder?(',' colNamecolOrder?)*
    query: SELECT expression (',' expression)* FROM srcorderBy

Chukwa

Chukwa is an open source data collection system for monitoring large distributed systems. Chukwa is built on top of the Hadoop distributed filesystem (HDFS) and MapReduce framework. There is a flexible and powerful toolkit for displaying, monitoring, and analyzing results to make the best use of the collected data available in Chukwa.

Flume

Flume is a distributed, reliable, and available service for efficiently collecting, aggregating, and moving large amounts of log data. It has a simple and flexible architecture based on streaming data flows. It is robust and fault tolerant with tunable reliability mechanisms and many failover and recovery mechanisms. It uses a simple extensible data model that allows for online analytic application.

Oozie

Oozie is a workflow/coordination system to manage Apache Hadoop jobs. Oozie workflow jobs are Directed Acyclical Graphs (DAGs) of actions like a MapReduce model. Oozie coordinator jobs are recurrent Oozie workflow jobs triggered by time (frequency) and data availability. Oozie is integrated with the rest of the Hadoop stack supporting several types of Hadoop jobs out of the box (Java MapReduce, Streaming MapReduce, Pig, DistCp, etc.) Oozie is a scalable, reliable and extensible system.

HCatalog

A new integrated metadata layer called HCatalog has been added to the Hadoop ecosystem recently (late 2011). It is built on top of the Hive metastore currently and incorporates components from Hive DDL. HCatalog provides read and write interfaces for Pig, MapReduce, and Hive in one integrated repository. By an integrated repository the users can explore any data across Hadoop using the tools built on its platform.

HCatalog's abstraction presents users with a relational view of data in the Hadoop distributed filesystem (HDFS) and ensures that users need not worry about where or in what format their data is stored. HCatalog currently supports reading and writing files in any format for which a SerDe can be written. By default, HCatalog supports RCFile, CSV, JSON, and sequence file formats, which is supported out of the box. To use a custom format, you must provide the InputFormat, OutputFormat, and SerDe, and the format will be implemented as it can be in the current Hadoop ecosystem. (For further details on HCatalog, please check with Apache Foundation page or HortonWorks).

Sqoop

As Hadoop ecosystem evolves, we will find the need to integrate data from other existing "enterprise" data platforms including the data warehouse, metadata engines, enterprise systems (ERP, SCM), and transactional systems. All of this data cannot be moved to Hadoop as their nature of small volumes, low latency, and computations are not oriented for Hadoop workloads. To provide a connection between Hadoop and the RDBMS platforms, Sqoop has been developed as the connector. There are two versions of Sqoop1 and Sqoop2, lets us take a quick look at this technology.

Sqoop1—In the first release of Sqoop, the design goals were very simple

- Export/Import data from enterprise data warehouse, relational databases, and NoSQL databases
- Connector-based architecture with plugins from vendors
- No metadata store

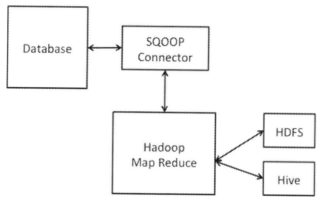

FIGURE 2.16 Sqoop1 architecture.

- Use Hive and HDFS for data processing
- Use Oozie for scheduling and managing jobs.

Installing Sqoop—currently you can download and install Sqoop from Apache Foundation homepage or from any Hadoop distribution. The installation is manual and needs configuration steps to be followed without any miss (Fig. 2.16).

Sqoop is completely driven by the client side installation and heavily depends on JDBC technology as the first release of Sqoop was developed in Java. In this workflow shown in Fig. 2.17, you can import and export the data from any database with simple commands that you can execute from a command line interface (CLI), for example.

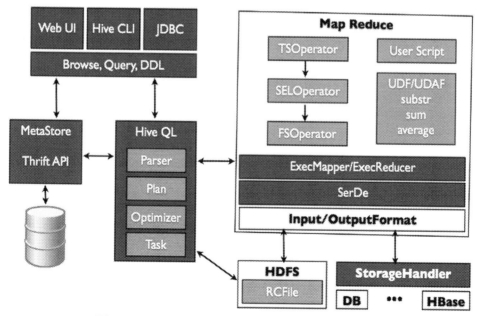

FIGURE 2.17 Hive process flow. *Image source—HUG discussions.*

Import syntax—sqoop import –connect jdbc:mysql://localhost/testdb \–table PERSON –username test –password ****.

- This command will generate a series of tasks
  - Generate SQL code
  - Execute SQL code
  - Generate maps/reduces jobs
  - Execute MapReduce jobs
  - Transfer data to local files or HDFS

Export syntax—sqoop export –connect jdbc:mysql://localhost/testdb \ –table CLIENTS_INTG –username test –password **** \ –export-dir/user/localadmin/CLIENTS

- This command will generate a series of tasks
  - Generate MapReduce jobs
  - Execute MapReduce jobs
  - Transfer data from local files or HDFS
  - Compile SQL code
  - Create or insert into CLIENTS_INTO table

There are many features of Sqoop1 that are easy to learn and implement, on the command line you can specify if the import is directly to Hive, HDFS, or HBASE. There are direct connectors to the most popular databases Oracle, SQL Server, MySQL, Teradata, and PostgreSQL.

There are evolving challenges with Sqoop1 including the following:

- Cryptic command line arguments
- Nonsecure connectivity—security risk
- No metadata repository—limited reuse
- Program driven installation and management

Sqoop2 is the next generation of data transfer architecture that is designed to solve the limitations of Sqoop1 namely (Fig. 2.18).

- Sqoop2 has a web-enabled UI
- Sqooptwo will be driven by a Sqoop Server architecture
- Sqoop2 will provide greater connector flexibility, apart from JDBC many native connectivity options can be customized by providers
- Sqoop2 will have a REST API interface
- Sqoop2 will have its own metadata store
- Sqoop2 will add credentials management capabilities, this will provide trusted connection capabilities

The proposed architecture of Sqoop2 is shown in Fig. 2.19. For more information on Sqoop status and issues please see the Apache Foundation website.

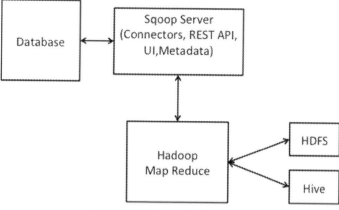

FIGURE 2.18 Sqoop2 architecture.

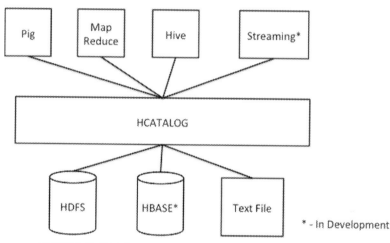

FIGURE 2.19 HCatalog concept. *Image original source—HortonWorks.*

NoSQL

Relational databases cannot handle the scalability requirements of large volumes of transactional data, and often fail when trying to scale up and scale out. The vendors of the RDBMS-based technologies have tried hard to address the scalability problem by replication, distributed processing and many other models, but the relational architecture and the ACID properties of the RDBMS have been a hindrance in accomplishing the performance requirements of applications such as sensor networks, web applications and trading platforms, and much more. In the late 1980s there were a number of research papers that were published about newer models of SQL databases, but not based on ACID requirements and the Relational model. Fast forward to 1998, the emergence of a new class of databases that can support the requirements of high-speed

FIGURE 2.20 Tipping point in NoSQL.

data in a pseudo database environment but nor oriented completely toward SQL were being discussed and the name NoSQL (Not only SQL) database was coined by Eric Evans for the user group meeting to discuss the need for nonrelational and nonSQL driven databases. This name has become the industry-adopted name for a class of databases, which work on similar architectures, but purpose built for different workloads.

There were three significant papers that changed the NoSQL database from being a niche solution to become an alternative platform (Fig. 2.20).

- Google publishes the Bigtable architecture (http://labs.google.com/papers/bigtable. html)
- Eric Brewer discusses the CAP Theorem
- Amazon publishes Dynamo (http://www.allthingsdistributed.com/2007/10/ amazons_dynamo.html)

Dynamo presented a highly available key–value store infrastructure and Bigtable presented a data storage model based on multidimensional sorted map, where a three-dimensional intersection between a rowkey, column key, and timestamp provide access to any data in petabytes of data. Both these scalable architectures had concepts where a distributed data processing system can be deployed at large scale to process different pieces of workload, with replication for redundancy and computations being driven programmatically. Both of these papers are the basis for further evolution of the architecture into multiple classes of databases. These architectures in conjunction with CAP theorem will be a discussion in the later sections of this book, when we talk about architecture of the future data warehouse and next generation analytics.

CAP theorem

Eric Brewer in the year 2000 presented a theory that he had been working for a few years at UC Berkley and at his company Inktomi, at the "Symposium on Principles of Distributed Computing". He presented the concept that three core systemic requirements that need to be considered when it comes to designing and deploying applications in a distributed environment and further stated the relationship between these requirements will create shear in terms of which requirement can you give up to accomplish the scalability requirements of your situation. The three requirements are: ***consistency***, ***availability,*** and ***partition tolerance***, giving Brewer's Theorem its other name—**CAP** (Fig. 2.21).

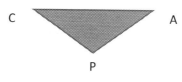

FIGURE 2.21 CAP theorem.

In simple terms CAP theorem states that in a distributed data system, you can guarantee two of the three requirements consistency (all data available at all nodes or systems), availability (every request will get a response) and partition tolerance (system will operate irrespective of availability or a partition or loss of data or communication). The system architected on this model will be called BASE (basically available soft state eventually consistent) architecture as opposed to ACID.

Combining the principles of the CAP theorem and the data architecture of Bigtable or Dynamo there are several solutions that have evolved—HBase, MongoDB, Riak, Voldemort, Neo4J, Cassandra, Hypertable, HyperGraphDB, Memcached, Tokyo Cabinet, Redis, CouchDB, and more niche solutions. Of these the most popular and widely distributed are the following:

- HBASE, Hypertable, Bigtable—architected on CP (from CAP)
- Cassandra, Dynamo, Voldemort—architected on AP (from CAP)

Broadly NoSQL databases have been classified into four subcategories.

Key–values pair—This model is implemented using a hash table where there is a unique key and a pointer to a particular item of data creating a key–value pair. Example—Voldemort andRiak

Column family stores—An extension of the key–value architecture with columns and column families, the overall goal was to process distributed data over a pool of infrastructure. Example—HBase and Cassandra.

Document databases—this class of databases is modeled after Lotus Notes and similar to key–value stores. The data is stored as a document and is represented in JSON or XML formats. The biggest design feature is the flexibility to list multiple levels of key–value pairs. Example—CouchDB.

Graph databases—Based on the graph theory, this class of database supports the scalability across a cluster of machines. The complexity of representation for extremely complex sets of documents is evolving. Example—Neo4J.

Let us focus on the different classes of NoSQL databases and understand their technology approaches. We have already discussed HBASE as part of Hadoopsections in this chapter.

Key–value pair—Voldemort

Voldemort is a project that originated in LinkedIn. The underlying need at LinkedIn was a highly scalable lightweight database that can work without the rigidness of ACID compliance. Dynamo and Memcached inspired the database architecture. Data is stored as a key with values in conjunction as a pair. Data is organized in a ring topology with redundancy and range management built into each node of the ring. The architecture is very niche in solving problems and hence did not get wide adoption outside of LinkedIn. It is still being evolved and updated at this time of writing.

Cassandra

Facebook in the initial years had used a leading commercial database solution for their internal architecture in conjunction with some Hadoop. Eventually the tsunami of users led the company to start thinking in terms of unlimited scalability and focus on availability and distribution. The nature of the data and its producers and consumers did not mandate consistency but needed unlimited availability and scalable performance. The team at Facebook built an architecture that combines the data model approaches of Bigtable and the infrastructure approaches of Dynamo with scalability and performance capabilities named Cassandra. Often referred as hybrid architecture as it combines the column-oriented data model from Bigtable with Hadoop MapReduce jobs and it implements the patterns from dynamo like eventually consistent, gossip protocols, a master—master way of serving both read and write requests. Cassandra supports a full replication model based on NoSQLarchitectures.

Cassandra team had a few design goals to meet, considering the architecture at the time of first development and deployment was primarily being deployed at Facebook. The goals included the following:

- High availability
- Eventual consistency
- Incremental scalability
- Optimistic replication
- Tunable tradeoffs between consistency, durability, and latency
- Low cost of ownership
- Minimal administration

Data model
Cassandra datamodel is based on a key—value model, where we have a key that uniquely identifies a value and this value can be structured or completely unstructured or can also be a collection of other key—value elements. This is very similar to pointers and linked lists in the world of programming. Fig. 2.22 shows the basic key—value structure.

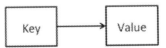

FIGURE 2.22 Key—value pair.

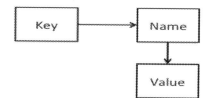

FIGURE 2.23 Cassandra key–value pair (column).

A key–value pair can represent a simple storage for Person→Name type of data but cannot scale much. An alteration to the basic model is done to create a name and value in the key–value pair, this would provide a structure to create multiple values and associate a key to the name–value pair. This creates a table like structure described in Fig. 2.23.

In the updated structure of the key–value notation, we can store Person → Name → John Doe, add another column called Person→Age→30 and create multiple storage structures. This defines the most basic structure in the Cassandra data model called "column".

Column—is an ordered list of values stored as a name–value pair. It is composed of a column name, a column value, and a third element called timestamp. The timestamp is used to manage conflict resolution on the server, when there is a conflicting list of values or columns to be managed. The client sets the timestamp that is stored along with the data, and this is an explicit operation.

The column can hold any type of data in this model, varying from characters to GUID to blobs. Columns can be grouped into a row called as rowkey. A simple column by itself limits the values you can represent, to add more flexibility, a group of columns belonging to a key can be stored together called as a column family. A column family can be loosely compared to a table in the database comparison.

Column Family—is a logical and physical grouping of a set of columns, which can be represented by a single key. The flexibility of column family is that the names of columns can vary from a row to another and the number of columns can vary over a period of time (Fig. 2.24).

There is no limitation with creating different column structures in a column family, except the maintenance of the same is dependent on the application that is creating the different structures. Conceptually it is similar to overloading in the object-oriented programming language.

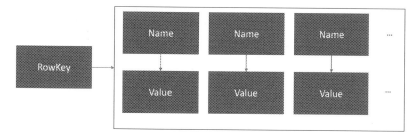

FIGURE 2.24 A column family representation.

If we wanted to further group column families together to create or manage the relationship between the column families, the Cassandra model provides a super column family.

Super Column Family—is a logical and physical grouping of column families that can be represented by a single key. The flexibility of this model is you can represent relationships, hierarchies, and tree-like traversal in a simple and flexible manner.

In order to create a meaningful data structure or architecture, a column family or super column family or multiples of the same need to be grouped in one set or under a common key. In Cassandra, a keyspace defines that set of column families grouped under one key. Typically we can decompose this as follows.

Excel Document → Sheet 1→ columns/Formulas→sheet2(columns/formulas)→ sheet 2(other columns/formulas) and so on. You can define a keyspace for an application, this is a preferred approach rather than create thousands of keyspaces for an application.

A keyspace has configurable properties that are critical to understand

- Replication factor—refers to the number of nodes that can be copies or replicas for each row of data. If your replication factor is 2, then two nodes will have copies of each row. Data replication is transparent. Replication factor is the method of controlling consistency within Cassandra and is a tunable parameter in deciding performance and scalability balance.
- Replica placement strategy—refers to how the replicas will be placed in the deployment (ring—we will discuss this in the architecture section). There are two strategies provided to configure which node will get copies of which keys. These are SimpleStrategy (defined in the keyspace creation) and Network Topology Strategy (replications across datacenters).
- Column families—Each keyspace has at least one or more column families. Column family has configurable parameters described in Fig. 2.25

Parameter	Default Value
column_type	Standard
compaction_strategy	SizeTieredCompactionStrategy
comparator	BytesType
compare_subcolumns_with	BytesType
dc_local_read_repair_chance	0
gc_grace_seconds	864000 (10 days)
keys_cached	200000
max_compaction_threshold	32
min_compaction_threshold	4
read_repair_chance	0.1 or 1 (See description below.)
replicate_on_write	TRUE
rows_cached	0 (disabled by default)

FIGURE 2.25 Column family parameters (for details see Apache or Datastax website).

As we have learned so far, a keyspace provides the data structure for Cassandra to store the column families and the subgroups. To store the keyspace and the metadata associated with it, Cassandra provides the architecture of a cluster, often referred as the "ring". Cassandra distributes data to the nodes by arranging them in a ring that forms the cluster.

Data partitioning

Data partitioning can be done either by the client library or by any node of the cluster and can be calculated using different algorithms; there are two native algorithms that are provided with Cassandra:

- The first algorithm is the RandomPartitioner—a hash-based distribution, where the keys are more equally partitioned across the different nodes, providing better load balancing. In this partitioning each row and all the columns associated with the rowkey are stored on the same physical node and columns are sorted based on their name.
- The second algorithm is the OrderPreservingPartitioner—creates partitions based on the key and data grouped by keys, which will boost performance of range queries since the query will need to hit lesser number of nodes to get all the ranges of data

Data sorting

When defining a column, you can specify how the columns will be sorted when results are returned to the client. Columns are sorted by the "compare with" type defined on their enclosing column family. You can specify a custom sort order, the default provided options are as follows:

- BytesType—Simple sort by byte value. No validation is performed.
- AsciiType—Similar to BytesType but validates that the input can be parsed as US-ASCII.
- UTF8Type—A string encoded as UTF8
- LongType—A 64-bit long
- LexicalUUIDType—A 128bitUUID, compared lexically (by byte value)
- TimeUUIDType: A 128bit version 1UUID, compared by timestamp
- Integer—Faster than a log, supports fewer or longer lengths.

Consistency management

The architecture model for Cassandra is AP with eventual consistency. Cassandra's consistency is measured by how recent and concurrent are all replicas for one row of data. Though the database is built on eventual consistency model, real world applications will mandate consistency for all read and write operations. In order to manage the user interaction and keep the consistency, Cassandra provides a model called tunable consistency. In this model, the client application decides the level of consistency desired by that application. This allows the user the flexibility to manage different classes of applications at different levels of consistency. There are additional built-in repair mechanisms for consistency management and tuning. A key point to remember is consistency depends on replication factor implementation in Cassandra.

Write consistency

Since consistency is a configuration setting in Cassandra, a write operation can specify its desired level of consistency. Cassandra lets you choose between weak and strong consistency levels. The following consistency levels are available.

Consistency level	Write consistency	Read consistency
ANY	A write must be written to at least one node If a replica is down, a live replica or current node can store a hint and update the node when it comes back live If all replica nodes for the given rowkey are down, the write can still succeed once has been written by storing the hint and the data in the coordinator Not a preferred model *Note—if all replica nodes are down an ANY write will not be readable until the replica nodes for that rowkey have restored*	Not applicable
ONE	A write must be written to the commit log and memory table of at least one replica node	Returns a response from the closest replica
QUORUM	A quorum is defined as the minimum number of replicas that need to be available for a successful read or write Quorum is calculated as (rounded down to a whole number): (replication_factor/2) + 1 For example, with a replication factor of 5, a quorum is 3 (can tolerate 2 replicas down) A quorum is a middle ground between weak and strong consistency A write must be written to the commit log and memory table on a quorum of replica nodes to be successful Local quorum—a write must be written to the commit log and memory table on a quorum of replica nodes in the *same* data center as the coordinator node Each quorum—a write must be written to the commit log and memory table on a quorum of replica nodes in *all* data centers	
ALL	A write must be written to the commit log and memory table on all replica nodes in the cluster for that row key The highest and strongest level of consistency brings latencies in the architecture Not a preferred method until complexity and volumes are low	Data is returned once all replicas have responded. The read operation will fail if a replica does not respond.

Read consistency

Read consistency level specifies how many replicas must respond before a result is returned to the client application. When a read request is made, Cassandra checks the specified number of replicas for the most recent data based on the timestamp data, to satisfy the read request.

Note—Localand Each Quorum are defined in large multi data center configurations.

Specifying client consistency levels

Consistency level is specified by the client application when a read or write request is made. For example,

SELECT *FROM CUSTOMERS WHERE STATE = 'IL' USING CONSISTENCY QUORUM;

Built-in consistency repair features

Cassandra has a number of built-in repair features to ensure that data remains consistent across replicas:

- **Read repair**—is a technique that ensures that all nodes in a cluster are synchronized with the latest version of data. When Cassandra detects that several nodes in the cluster are out of sync, it marks the nodes with a Read Repair flag. This triggers a process of synchronizing the stale nodes with newest version of the data requested. The check for inconsistent data is implemented by comparing the clock value of the data and the clock value of the newest data. Any node with a clock value that is older than the newest data is effectively flagged as stale.
- **Antientropynode repair**—is a process that is run as a part of maintenance and called as Nodetool process. This is a synchronized operation across the entire cluster where the nodes are updated to be consistent. It is not an automatic process and needs manual intervention. During this process, the node exchange has information represented as "Merkletrees," and if the tree information is not consistent, a reconciliation exercise needs to be carried out. This feature comes from Amazon Dynamo, with the difference being; in Cassandra each column family maintains its own Merkle Tree.

A quick note, a Merkle tree is a hash key hierarchy verification and authentication technique. When replicas are down for extended periods, the Merkle tree keeps checking small portions of the replicas till the sync is broken, enabling a quick recovery (or information on Merkle Trees, check Ralph Merkle's webpage www.merkle.com)

- **Hinted handoff**—During a write operation, data is set to all replicas by default. If a node is down at that time, data is stored as a hint to be repaired when the node comes back. If all nodes are down in a replica, the hint and the data are stored in the coordinator. This process is called as a hinted handoff. No operation is permitted in the node until all nodes are restored and synchronized.

Cassandra ring architecture

Fig. 2.26 shows the ring architecture we described earlier. In this configuration, we can visualize how Cassandra provides for scalability and consistency.

In the ring architecture, the key is the connector to the different nodes in the ring, and the nodes are replicas. For example, A can be replicated to B and C, when N = 3. And D can be replicated to D and E or D and F when N = 2.

Data placement

Data placement around the ring is not fixed in any default configuration. Cassandra provides two components called snitches and strategies to determine which nodes will receive copies of data.

- Snitches define proximity of nodes within the ring and provides information on the network topology
- Strategies use the information snitches provide them about node proximity along with an implemented algorithm to collect nodes that will receive writes.

Data partitioning

Data is distributed across the nodes by using "partitioners". Since Cassandra is based on a ring topology or architecture, the ring is divided into ranges equal to the number of nodes, where each node can be responsible for one or more ranges of the data. When a node is joined to a ring, a token is issues and this token determines the node's position

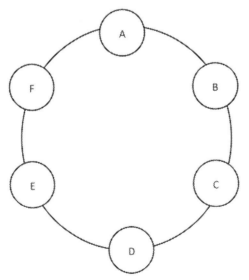

FIGURE 2.26 Cassandra ring architecture.

on the ring and assigns the range of data it is responsible for. Once the assignment is done, we cannot undo it without reloading all the data.

Cassandra provides native partitioners and supports any user-defined partitioner. The key feature difference in the native partitioner is the order preservation of keys.

Random Partitioner—is the default choice for Cassandra. It uses an MD5 hash function to map keys into tokens, which will evenly distribute across the clusters. Random partitioning hashing techniques ensures that when nodes are added to the cluster, the least possible set of data is affected. While the keys are evenly distributed, there is no ordering of the data, which will need the query to be processed by all nodes in an operation.

Ordered preserving partitioners—as the name suggests, preserves the order of the row keys as they are mapped into the token space. Since the key is placed based on ordered list of values, we can run efficient range–based data retrieval techniques. The biggest drawback in this design is a node with its replicas may become unstable over time especially with large reads or writes being done in one node.

Peer to peer—simple scalability

Cassandra by design is a peer to peer model of architecture, meaning in its configuration there are no designated master or slave nodes. The simplicity of this design allows nodes to be taken down from a cluster or added to a cluster with ease. When a node is down, the processing is taken over by the replicas and allows for a graceful shutdown, similarly when a node is added to a cluster, upon being designated with its keys and tokens, the node will join the cluster and understand the topology before commencing operations.

Gossip protocol—node management

In Cassandra architecture, to manage partition tolerance and decentralization of data, managing intranode communication becomes a key feature. This is accomplished by using the gossip protocol. Alan Demers, a researcher at Xerox's Palo Alto Research Center, who was studying ways to route information through unreliable networks, originally coined the term "gossip protocol" in 1987.

In Cassandra, the gossip protocol is implemented as gossiper class. When a node is added to the cluster it also registers with the gossiper to receive communication. The gossiper selects a random node and checks it for being alive or dead, by sending messages to the node. If a node is found to be unresponsive, the gossiper class triggers the "hinted handoff" process if configured. In order for the gossiper class to distinguish between failure detection and long running transactions, Cassandra implements another algorithm called "Phi Accrual Failure Detection algorithm" (based on the popular paper by Naohiro Hayashibara et al.). According to the "accrual detection" algorithm, a node can be marked as suspicious based on the time it takes to respond and more the delays, higher the suspicion that the node is dead. This delay or accrued value is calculated by the Phi and compared to a threshold, which will be used by the gossiper to determine the state of the node. The implementation is accomplished by the "failuredetector" class, which has three methods:

- isAlive(node_address)—What the detector will report about a given node's aliveness.
- interpret(node_address)—this method is used by the gossiper to make a decision on the health of the node, based on the suspicion level reached by calculating Phi (the accrued value of the state of responsiveness)
- report(node_address)—When a node receives a heartbeat, it invokes this method.

With the Peer to Peer and gossip protocols implementation, we can see how the Cassandra architecture keeps the nodes synced and the operations on the nodes scalable and reliable. This model is derived and enhanced from Amazon's Dynamo paper. Based on the discussion of Cassandra so far, we can see how the integration of two architectures from Bigtable and Dynamo has created a row-oriented column-store, that can scale and sustain performance. At this time of writing Cassandra is a top level project in Apache. Facebook has already moved on to proprietary techniques for large-scale data management, but there are several large and well-known companies that have adopted and implemented Cassandra for their architectural needs of large data management especially on the web, with continuous customer or user interactions.

There are a lot more details on implementing Cassandra and performance tuning, which will be covered in the latter half of this book when we discuss the implementation and integration architectures.

Basho Riak

Riak is a document oriented database. It is similar in architecture to Cassandra, and the default is setup as a four-node cluster. It follows the same ring topology and gossip protocols in the underpinning architecture. Each of the four nodes contains eight nodes or eight rings, thus providing a 32 ring partition for use. A process called vnodes(virtual nodes) manages the partitions across the 4 node cluster. Riak uses a language called erlang and MapReduce. Another interesting feature of Riak is concept of links and link walking. Links enable you to create metadata to connect objects. Once you create links, you can traverse the objects and this is the process of link walking. The flexibility of links allows you to determine dynamically how to connect multiple objects. More information on Riak is available at Basho's (the company that designed and developed Riak)website.

Other popular NoSQL implementations are document databases (CouchBase, MongoDB, and other) and Graph Databases (Neo4j). Let us understand the premise behind the document database and graph database architectures.

Document oriented databases or document database can be defined as a schema less and flexible model of storing data as documents, rather than relational structures. The document will contain all the data it needs to answer specific query questions. Benefits of this model include the following:

- Ability to store dynamic data in unstructured or semistructured or structured formats.
- Ability to create persisted views from a base document and storing the same for analysis
- Ability to store and process large data sets.

The design features of document-oriented databases include the following:

- **Schema free**—there is no restriction on the structure and format of how the data needs to be stored. This flexibility allows an evolving system to add more data and allows the existing data to be retained in the current structure.
- **Document store**—Objects can be serialized and stored in a document, there is no relational integrity to enforce and follow.
- **Ease of creation and maintenance**—A simple creation of the document allows complex objects to be created once and there is minimal maintenance once the document is created.
- **No relationship enforcement**—Documents are independent of each other and there is no foreign key relationship to worry when executing queries. The effects of concurrency and performance issues related to the same are not a bother here.
- **Open formats**—Documents are described using JSON or XML or some derivative, making the process standard and clean from the start.
- **Built-in versioning**—Documents can get large and messy with versions. To avoid conflicts and keep processing efficiencies, versioning is implemented by most solutions available today.

Document databases express the data as files in JSON or XML formats. This allows the same document to be parsed for multiple contexts and the results scrapped and added to the next iteration of the database data.

Example usage—A document database can be used to store the results of clicks on the web. For each log file that is parsed a simple XML construct with the Page_Name, Position_Coordinates, Clicks, Keywords, Incoming and Outgoing site and date_time will create a simple model to query the number of clicks, keywords, date, and links. This processing power cannot be found in an RDBMS. If you want to expand and capture the URL data, the next version can add the field.

The emergence of document databases is still ongoing at the time of this book (2012) and the market adoption for this technology will happen soon. We will discuss the integration architecture for this technology in the latter half of this book.

Graph databases

Social media and the emergence of Facebook, LinkedIn, and Twitter have accelerated the emergence of the most complex NoSQL database, the graph database. The graph database is oriented toward modeling and deploying data that is graphical by construct. For example—to represent a person and their friends in a social network, we can either write code to convert the social graph into key–value pairs on a Dynamo or Cassandra or simply convert them into a node-edge model in a graph database, where managing the relationship representation is much more simplified.

A graph database represents each object as a node and the relationships as an edge, this means person is a node and household is a node, the relationship between the two is an edge.

Data model—like the classic ER model for RDBMS, we need to create anattribute model for a graph database. We can start by taking the highest level in a hierarchy as a root node (akin to an Entity) and connect each attribute as its subnode. To represent different levels of the hierarchy we can add a subcategory or subreference and create another list of attributes at that level. This creates a natural traversal model like a tree traversal, which is similar to traversing a graph. Depending on the cyclic property of the graph, we can have a balanced or skewed model. Some of the most evolved graph databases include Neo4J, infiniteGraph, GraphDB, and AllegroGraph.

There are additional Hadoopcommitters and distributors like MapR and these architectures will be covered in the Appendix.

AS we conclude this chapter we see the different technologies that are available to process big data, their specific capabilities, and their architectures. In the next chapter we will study some use cases from real life implementations of solutions. In the second half of this book we will see how these technologies will enrich the data warehouse and data management with big data integration.

For continued reading on specific vendors for NoSQL databases, please check their websites.

Additional reading

Hive A Data Warehousing Solution Over a MapReduce Framework - Facebook Data Infrastructure Team.

Apache Software Foundation Page.

Pavlo et al. A Comparison of Approaches to Large-Scale Data Analysis. Proc. ACM SIGMOD, 2009.

C. Ronnie et al. SCOPE: Easy and Efficient Parallel Processing of Massive Data Sets. Proc. VLDB Endowement 1(2):1265–1276, 2008.

J. Dean and S. Ghemawat. MapReduce: a data processing tool. Communications of the ACM, 53(1): 72–77, 2010.

D.J. DeWitt and M. Stonebraker. MapReduce: A major step backwards. The Database Column, 1, 2008.

E. Friedman, P. Pawlowski, and J. Cieslewicz. SQL/MapReduce: A practical approach to self-describing, polymorphic, and parallelizable user-defined functions. Proceedings of the VLDB Endowment, 2(2): 1402–1413, 2009.

S. Ghemawat, H. Gobio, and S. Leung. The Google File system. SIGOPSOper. Syst. Rev., 37:29:43, October 2003.

A. Thusoo, J.S. Sarma, N. Jain, Z. Shao, P. Chakka, S. Anthony, H. Liu, P. Wycko, and R. Murthy. Hive: a warehousing solution over a Map-Reduce framework. Proceedings of the VLDB Endowment, 2(2): 1626–1629, 2009.

Andrew Pavlo, E. Paulson, A. Rasin, D.J. Abadi, D.J. Dewitt, S. Madden, and M. Stonebraker. A Comparison of Approaches to Large-Scale Data Analysis. Brown University. Retrieved 2010-01-11.

Colby Ranger; Ramanan Raghuraman, Arun Penmetsa, Gary Bradski, and Christos Kozyrakis. "Evaluating MapReduce for Multi-core and Multiprocessor Systems". HPCA 2007, Best Paper.

Hadoop Sort Benchmarks, Owen O'Malley and Arun C. Murthy, 2009.

A Comparison of Join Algorithms for Log Processing in MapReduce (S. Blanas, J. Patel, V. Ercegovac, J. Rao, E. Shekita, Y. Tian).

Cassandra - AStructured Storage System, Avinash Lakshman,Prashant Malik Facebook.

The CAP Theorem by Seth Gilbert & Nancy Lynch.

Time, Clocks, and the Ordering of Events in a Distributed System by Leslie Lamport.

3

Building big data applications

Human language is the new UI layer, bots are like new applications, and digital assistants are meta apps. Intelligence is infused into all of your interactions

Satya Nadella

Data storyboard

Information floats around us nonstop. It is a fascinating layer of details that can be harnessed, integrated, analyzed, repeated for play purposes, implement metrics, and move to next steps. This layer of details contains data which is structured, semi-structured, and unstructured, all of which can be often connected to one user or multiple users of one group, or even multiple groups. The fascination to study and interpret different patterns of behavior and hidden insights that can lead to wonderful business outcomes has always been a dream for any enterprise or individual in the world. We constantly see many examples of this today with the virtual connected world we live in, but the same patterns have existed since the stone age of human existence, and we have read in history how many times our minds have innovated ideas for trade and have created situations of advantage to one individual or group, over the others. The bottom-line today is the availability of all tipping points in one unison to make this possible; however, it requires a methodology, the ability to box and isolate all the raw noise and interpret the needed signals, the ability to scale up and scale out on demand, the processing of all kinds of data in own format and integration into one useable format, and the requirements of neural networks and machine learning to implement backend processing and automation of tasks. Let us take a deeper look into all these areas with examples.

Information Layers—a multi layered series of data points that are interconnected loosely and can be harnessed to create a lineage and join the different touchpoints of granularity efficiently and produce the outcomes. However, the issue here is the fact that we do not harness all the raw data today and hence there is a generation of data consumers who do not know the intricacies of this approach and the lower level methods of integration. An example of the layers is discussed below.

As seen in Fig. 3.1, there are interactions that I call as "market interactions" and these are all events that occur, in parallel at multiple locations and can be tethered to deliver outcomes which are delivered as analytics. The foundation of our lives has always been driven by the usage of analytics based on the time of its delivery. Whether it is history or space exploration, we are constantly guided by the events and outcomes. There are

Building Big Data Applications. https://doi.org/10.1016/B978-0-12-815746-6.00003-X

73

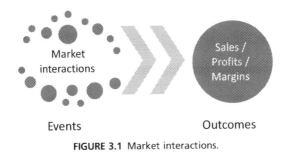

Events   Outcomes

FIGURE 3.1 Market interactions.

several interactions happening in the events layer and these can be simple or complex, and will need processing within the layer of production as soon as they occur.

The layers of information will contain different types of data and they need to be managed with accuracy and clarity, which will require classification, categorization, segmentation, formulation, association, metadata, master data, and details ascertained. Let us proceed with this situation to see how a market or farmer's market scenario would have worked in the past or even in today's world next.

Fig. 3.2 discusses the steps of a market or farmer's market and its activities. In the past or even today these markets happen in the weekend and typically on a Sunday. The market opens around 9:00a.m. and closes around 4:00p.m., with all types of goods being sold from groceries, fish, meat, garage sales of electronics, records, tapes, musical instruments, older versions of televisions, video cassette recorders, blue-ray players and more; stores selling sunglasses, clothing, and miscellaneous items. The people who shop here include all ages and socio-economic status as some of the vendors have a reputation that has been earned selling specific items that are excellent in quality and their

Market Opens on Saturday
Vendors arrive and setup their stalls
Prospects arrive and stroll across the market
Certain prospects turn into customers based on vendor good and prices
There are regular customers that come and shop for goods as normal
Conversations and chatter around the market provide inputs to prospects
The word of mouth discussion leads to certain vendors profiting more than others
There is also the creation of customers from these discussions that mean more crowds at next week
These crowds form communities that become the brand creators and torch bearers
The events discussed here are the traditional market happenings, which can be experienced today in farmers markets if you need to see this in action

FIGURE 3.2 Farmers markets interactions.

capability to deliver the best every week in terms of the product. In terms of how these vendors have got this branding is where we need to pay attention to the process.

The interactions of the vendors with customers from the first visit when the prospect arrived at their store to customer conversion, acquisition and follow-up conversations week over week, led to whispers between friends who meet at the market leading to more referral of customers, which explodes as the friends who were introduced lead more friends and the pattern continues to evolve. This continuum is where the word of mouth marketing plays an excellent role.

The vendors who acquire this popularity also start running their statistics of sales increase and directly relate the same to newer customers. This leads to better product procurement at lower price points for their inventory which is sold at a good profit margin while maintaining a healthy competitive price management. The market style of management of customers includes statistics of profit management, pricing of inventory from procurement to sale, the overall calculation of storage of the product, and the efficiency of managing the location to ensure all customers get their goods and are serviced.

This entire process tells us three things are important:

- *The data about the customers*
- *The data about the product*
- *The analytics of the sales*

In addition to these things, there are several aspects that are needed to manage and execute the following:

- *Collection of data*
- *Governance of the sales*
- *Governance of the customers*
- *Governance of the product*
- *Management of word of mouth marketing*
- *Collection of customer requirements on a weekly basis*
- *Solicitation of feedback*
- *Management of complaints*

Let us look at the next sets of information layers before we discuss the process items.

Mall Interactions—Let us look at the life of customers as we opened up malls and supermarkets. The customer here is introduced to a concept of self-service of all items they wish to buy, and even payment is managed by machines in many cases. In this situation the complexity of understanding the customer and the prospect is very deep and requires a high degree of efficiency and ease of management and decision support.

The mall activities happen in this order, the business opens every day at 9:00 a.m. and vendors open at the same time for all types of business including clothing, appliances, food, electronics, mobile phone services, specialty stores, and more. The prospective shoppers including customers plan a visit to the mall and often shop during Friday evenings, Saturday, and Sunday all day. The core issue here is nobody knows when these customers and prospects will come and what they will purchase. Who has been promoted in a campaign, what percentage of discounts have been offered, what all vendors are running the campaign. In the 1980s to early 2000s the businesses in the malls were running as there was no other segment to attract the purchase of the customers. With this situation, all vendors had customers and even if there were crossover of customers, the exchange of the customers across the segments became profitable.

The customers and prospects entertain chatter between them and often these communities that are formed become a word of mouth group for vendors, the special products they sell and the best price points that have been experienced by these customers (Fig. 3.3).

There are several lessons learned from this mall interactions segment. This entire process tells us several things are important:

- *Customers and prospects are similar in behavior*
- *Products can be sold in different segments by different vendors at different prices*
- *Campaigns can be executed at any interval by any vendor*
- *Different specialty stores will open up for every season*
- *Profitability needs to be managed for all segments and costs*
- *Returns must be accepted for a time period by the vendors*
- *Customers can return any product within the specific period for a full refund*
- *The analytics of the sales, customers, products, inventory, employees, associates, and location will be available in 48 h approximately*

In addition to these things, there are several aspects that are needed to manage and execute:

- *Collection of data of all types and across the location for all segments*
- *Governance of the sales*

Mall opens for business
Vendors open their stores
Prospects arrive and stroll across the mall
Certain prospects turn into customers based on vendor good and prices
There are regular customers that come and shop for goods as normal
Conversations and chatter around the mall provide inputs to prospects
The word of mouth discussion leads to certain vendors profiting more than others
There is also the creation of customers from these discussions that mean more crowds at next week
These crowds form communities that become the brand creators and torch bearers
The events discussed here are the mall happenings, which can be experienced today if you need to see this in action

FIGURE 3.3 Mall interactions.

- *Governance of the customers*
- *Governance of the product*
- *Governance of the employees at the location*
- *Management of marketing and campaigns*
- *Solicitation of feedback*
- *Management of complaints*

The complexity of all of these processes is where the latency of the system and the platforms came into play. There are exorbitant cost models that need to be managed, supply chain needs to be optimized and managed, calendar and seasonal sales need to be studied and optimized, write-off and disposal of inventory need to be managed, and all compliance and regulatory management and financial reporting need to be submitted to the government agencies on time.

We have gone from collecting data in the systems called online transaction processing (OLTP) to defining and building the data warehouse (DW) and analytical databases (AD), while pursuing the elusive tipping point for computing all data at any time for any event across the platform. This is where the next stage of data evolution and more complexity to information layers can be introduced, also known as the Internet.

eCommerce Interactions *started when the new world of shopping and browsing was enabled with the Internet. We moved from the concept of markets and malls into a virtual world where you can now sell to prospects across the whole world. The concept of a store and keeping it open for business between a certain time period disappeared and instead we learned that the store is always open for business. This model provided a shock first to many as we needed to understand the concept of actually creating an online catalog that had the details of each and every product, the cost and any discounts offered, and all of this was visible to all the people meaning the competition and the suppliers. From a prospect viewpoint the actual beauty was the ability to shop across a variety of vendors, with different price models and promotions. This world driven by ecommerce caused a storm in the business opportunity* (Fig. 3.4).

We learned that we needed to have catalogs, prices, discounts, promotions, deals, strategies, collaborations, newer venture models, and most important of all the dealing with a totally new prospect. The new prospect is very different from the prior generation in the viewpoint that they will browse for products, prices, deals, and promotions at any time from their home or work or any place they will be, and the time of their browse and stay on your store is very critical as you have their attention for the few seconds. If the deal is locked in those few seconds you have a conversion and a customer else the data is now available for others as the cookies that are launched by all vendors keep sniffing every second for the product search from a prospect. This means you need to compete different.

The ecommerce world taught us the value of instant analytics, which means that we can measure the value delivered to any prospect who is looking at a product in the shortest duration of time. This series of analytics drove the first generation of machine

Internet is open 24x7x365 for business
Vendors are selling in a continuum
Prospects and Customers are there and looking for deals, and there is a social media buzz on deals and prices
Each vendor needs to be able to trace the steps of each prospect as they browse for items
Every millisecond of stay increases an opportunity to sell, there needs to be a just in time marketing that needs to happen
The prospect can be attracted by other vendors also communicating with cookies and offers, while they are on your site
The ability to convert from look to buy is the most interesting conversion and the point of victory for any vendor
The crowd that strolls, looks and converts to buy can form a community that provides the word of mouth marketing for a vendor
The events discussed here are the eCommerce happenings, which can be experienced today if you need to see this in action

FIGURE 3.4 eCommerce Interactions.

learning where the machine was taught to compute in the parallel domains how the behavior of the prospect was during the entire browse period. We developed a series of algorithms for the family of behavior which includes recommendation algorithms, sentiment analysis algorithms, influencer analytics algorithms, location behavior algorithms, and socio-economic algorithms.

The compute logic in the world of internet evolved from the first minute of its birth. We realized that using a database and a storage area network was good but cannot scale in terms of the volume of data, the velocity of its production and the variety of formats. We needed to have an ecosystem where the compute shall move to where the data lives; and this was proven by Google, Netscape, and America OnLine first and repeated by Yahoo, Amazon, Facebook, Twitter, and others soon. The new compute mechanism introduced distributed data processing as the foundational family of architecture and on this design and patterns wereborn a series of platforms including Apache Hadoop, Google Dremel, Google File System, Amazon DynamoDB, Apache Cassandra, Apache HBase, and other evolving platforms. A newer series of programming languages has also evolved in terms of Javascript, Python, Go, Scala, and others.

The ecommerce model has evolved into a mcommerce model with mobile platforms being a provider of information to any prospect at any time anywhere. This means we have now learned multiple lessons and strategies in terms of information management and analytics, these include the following:

- *Data collection expands from being just about what we had browsed to what is being searched. This means collection of clicks, links, pages, products, comparison, time, active connection and presence, and potential conversion opportunities. All of this data needs to be collected automatically and analyzed for next steps.*

- *Analytics needs to happen now and later. This requires models that can execute with speed, an acceptable margin of error, acceptable ranges of recommendations for discounts, and competitive offers. The analytics needed to be not geared to prescriptive but rather an introspective.*
- *What these changes to data and analytics drive us to implement is the change of how to market, where the generation that needs to be sold to today wants offers at the time of their shopping not before, which means you need to be beyond the thought of agile marketing. Here the concept to delivery of a campaign is in the duration of the browsing and is tailored to the customer preferences, which are collected based on sentiments and user behavior analysis, the socio-economic models of the zip codes where the customers reside. These factors require us to not execute segments and outliers but rather create buckets based on specific criteria and then validate which buckets align closest to the market at the time of execution. This complexity is very tough to resolve and takes companies time and effort. But this is trick to resolve to be successful.*
- *The process of complexity does not end here; we need to have the optimized supply chain process and inventory management process, which will ensure that when the customer places an order, the entire order will be shipped and delivered as promised to the customer on that date and not have delays. This means your store in the ecommerce model of business is connected with distribution centers which will have access to inventory and shippable information ready. The model further changes when you see back orders for a popular item and this means you will need to have creative marketing strategies and offers to induce the conversion to become an order. There are several complexities in integrating the data to make this happen seamlessly. Today when you look at Amazon or Baidu they have invested time, energy, and effort in creating these platforms. However, they also have a large workforce that is available to manage all the processes, and this is where we will see the evolution of process automation and integration.*
- *As we look at the governance mechanisms and strategies, the ecommerce world has added more transparency within the enterprise that requires us to ensure that*
  - All appropriate users can see data on demand
  - All appropriate users can execute analytics on demand
  - All data is secure
  - All customer information is secure and privacy is 100% enabled
  - All product details are disclosed as required for each sale
  - All complaints are handled by a 24 × 7 × 365 call center
  - All languages are supported in the call center
  - All data is digital
  - All processes are digital
  - All complexity is documented and formulated
  - All decisions are available to be replayed as needed to learn and train.

These lessons and learnings are now enabling more complexity to be handled, which we will discuss in the next two sections, one on healthcare and the other on research.

Patient care has been a big problem to solve for the governmental agencies, insurance companies, pharmacies, physicians, and patients. In a traditional method that was followed for many years, we as patients will setup an appointment with a physician and on the first visit to the office, fill information about ourselves, family history, reason for visit, insurance information, employment information, and if the visit is for a family member, the primary insured will sign a financial support section. All of this once complete, will provide the physician the ability to bill the insurance company for your visit and their time spent with you.

Each follow-up visit will provide opportunity to collect details of what is called an episode, which is your current state of health and associated encounters or reactions to medicine or diet or exercise. Each episode can have more than one encounter and each encounter can mean you are provided different medicines and prescriptions. Now if there were paper collection of all of these details, it became burdensome as time went by, and if people changed physicians for whatever reason, all of the paper had to be transferred to the new physician. This meant three issues:

- *Security of data*
- *Paperwork inefficiencies*
- *Loss of productive time*

As we evolved in the world of digital data, we started using Representational State Transfer (RESTful) application programming interfaces (APIs) to manage data exchange between applications and systems. To improve the data management of a patient we need to think of creating REST API layers that can deliver data in an interoperable layer (Fig. 3.5).

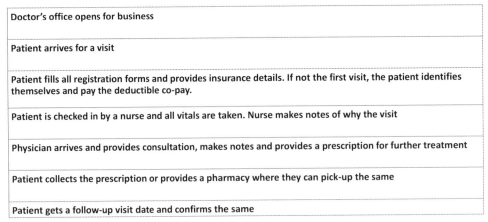

FIGURE 3.5 Doctor's office.

By changing the goal of patient care to be driven digital, we are emerging into the world of electronic healthcare records (EHR), which means the patient can deliver the management of their records more efficiently and provide the needed interoperability by sharing all the data between doctors as needed. There are governance policies that need to be implemented for this transformation to happen successfully.

There are layers of complexity, privacy, compliance and regulations, lineage, sourcing, tagging, metadata, and management rules that need to be implemented. To create more details of information processing there are associated data elements that need integration and appropriate metadata added. The outcome we are expecting is something like this.

In Fig. 3.6 we have shown the details of the people who are involved in providing healthcare, the data that is created in the healthcare and the process that are used to manage and deliver the required healthcare. The complexity is present across all the layers and in all the cases the most complex tasks are in the process components while the security layers between people and data are very complex to manage on a daily basis. The new age of healthcare will need this interoperability platform to be built and using the big data platform technologies along with REST API we can definitely create this. The details of this will be discussed in a later chapter.

As we see the information layers and its complexity, there is another use case to discuss where the actual complexity needs to be replayed multiple times to deliver the real value, it is the use case of research. Fig. 3.7 shows the use case and the process we follow in the research across labs. The labs across the different areas of research consist

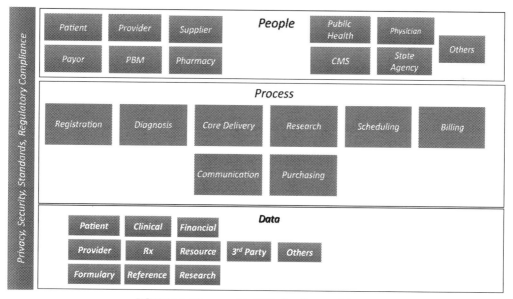

FIGURE 3.6 Interoperable EHR—healthcare of future.

Lab research happens 24x7x365
There are several teams of researchers that work on a program
They all conduct experiments constantly
They record all the steps that are taken in the process, the formulas, the outcomes at each stage, the models, the fitting of outcomes to the models, any errors and how they are handled
All the data is vital to understand when an error occurs and how to improve the efficiency to prevent the re-occurrence of the error
The core requirement of a platform here is the ability to record and replay the experiment as many times needed.
The replay will provide the immense insights that is needed.
A new form of integration and reuse of information with details at each layer and its outcomes will deliver big benefits

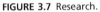

FIGURE 3.7 Research.

of team members who will work different times and execute different experiments. We need to ensure that all experiments are recorded as they progress with formulas, calculations, outcomes, errors, and any failures.

The experiments steps once recorded will be replayed as many times as needed, and we can figure out risks and issues as they occur. This experiment use case will provide more value as we understand the complexity in the process.

The process steps when recreated with integrating everybody's results will also provide an opportunity to ensure all risks are mitigated and the final execution run can be planned and experimented.

The core focus that we have understood is the need to define the information layer management and processing. This aspect is different in the new world of information as we start moving into different aspects is data management including DevOps and agile project management which includes Kanban and Scrum/XP methodologies of code development. The new world is information floating around us in a continuum, and the layers of data within this information can be simple to extremely complex. What do we do to be successful? The first order of business here is to isolate the noise and the value quotient in the information, which leads us to deliver analytics and performance indicators which are the requirements for business to work with data. This means two goals to accomplish, first the business needs to own the program and second the business needs to drive the program, from a governance perspective. The business subject matter experts will need to line up the rules and models of how the information layer will be accessed and used, who can access the raw layers of data, and who can operate the data for consumption by the larger team. The aspects of governance especially need to be succinct and clear as the noise versus value ratio is very complex to manage.

The new data layer needs to be harnessed with machine learning algorithms and an artificial intelligence data munging and processing layer. This layer is where the intellectual property and enterprise-specific analytic intelligence can be built and deployed. This layer also will be crucial in securing the enterprise story of information management as IT will don the role of the facilitator and not the owner of the processing layers which will mean budgetary shifts and controls in the entire processing engine of the enterprise. This is what I refer to as the impact of data science.

Data science is not new but rather has become a very important method to focus for delivery of big data processing and management of all applications related to big data. The concept further needs to be standardized as each enterprise is implementing its own version of a data science team. The role of a data scientist is to become an explorer, a researcher, an analytic nomad, an enthusiast of data discovery, and a deep sea diver. Each of these behaviors have traits associated with them and these traits will be the key to what the business wants to achieve in the world of information management and applications. These newer roles will also define the governance model of execution as business subject matter experts will not only define the rules but also enforce the rules successfully. In this model of governance is where the risks of big data application are managed effectively and succinctly.

The best-in-class companies that have experienced success have repeatedly said that shifting the gears of application building and management from the hands of IT has been critical to the overall success. While this concept is not new, the confusion of who owns the data in the enterprise and the stewardship of the data within the enterprise has emerged to be an intense battle of wits and courage. The biggest risk is not identifying the owners of the layers of information, and not defining the models of governance which will lead you to failures and disappointment with this new world of data and applications.

Further additions to data science have emerged in the world of information management with robotic process automation and low code models of user interface development. These new additions provide devOps teams benefits including ability to develop in Python for backend processing, the ability to add frameworks for simplifying the data collection and tagging exercises, the ability to create user interface layers in a nonmodel-view-controller architecture, all of which provide applications layers immense performance benefits.

No discussion is complete without touching security. In the new world of application processing, security takes a whole new level of presence in the enterprise. The applications will be managed as services with isolation layers built to shield the actual layers of data, which means even if there is a breach and compromises are made in the application layer it will be detected, isolated, and managed without the data layer getting affected and impacted. This is a discussion of how cyber security will work in the enterprise, and we will cover this in a chapter toward the end of this book.

As we conclude this chapter, let us look at the critical points we need to take into consideration for the new world of applications:

- *Governance is critical*
- *Stewardship is the biggest risk and pitfall to manage*
- *Automation is here and now*
- *Data science is the new method to success*
- *Business owns and manages data*
- *IT is the facilitator*
- *We looked at several examples and identified these areas as we discussed the examples*
- *Application processing requires us to move beyond just data and information*

As we move into the next segments of the book, we will look at how each segment is using these approaches to manage delivery of big data applications and its impact to that organization.

4

Scientific research applications and usage

This chapter focuses on the building of a big data application that is used within CERN, where the complexity of large data loads and its automated analysis and insights execution is the norm. This will include the discussion on data ingestion, large data sets, streaming data sets, data computations, distributed data processing, replications, stream versus batch analytics, analytic formulas, once versus repetitive executions of algorithms, supervised and unsupervised learning, neural networks execution, and applying visualizations and data storyboarding. The research community benefits of the big data stack and applications, how to avoid risks and pitfalls, when to boil the ocean, and where to avoid traps will all be discussed in the last segment of this chapter.

The name CERN is derived from the acronym for the French "Conseil Européen pour la Recherche Nucléaire," or European Council for Nuclear Research, a provisional body founded in 1952 with the mandate of establishing a world-class fundamental physics research organization in Europe. At that time, pure physics research concentrated on understanding the inside of the atom, hence the word "nuclear."

In 1964, there were two physicists François Englert and Peter Higgs who published individual papers on particle physics and defined that there must exist a particle that is smaller than an atom and even smaller than a proton and electron, which can be called as "God particle". To illustrate the name they called it a "boson" after Satyendranath Bose another physicist who collaborated with Albert Einstein on several experiments. Some of the pair's work resulted in the invention of Bose–Einstein statistics, a way to describe the behavior of a class of particles that now shares Bose's name. Two bosons with identical properties can be in the same place at the same time, but two fermions cannot. This is why photons, which are bosons, can travel together in concentrated laser beams. But electrons, which are fermions, must stay away from each other, which explains why electrons must reside in separate orbits in atoms. The boson discovery when done will open several new ways of understanding the universe which till date has been understood around 4%.

Building Big Data Applications. https://doi.org/10.1016/B978-0-12-815746-6.00004-1

The Higgs boson is the last undiscovered particle predicted by the Standard Model, a beautiful mathematical framework physicists use to describe the smallest bits of matter and how they interact. Experimental results have time and again validated the model's other predictions. But finding the Higgs boson would not close the book on particle physics. While the Standard Model accounts for fundamental forces such as electro-magnetism and the strong nuclear force, it cannot make sense of gravity, which is disproportionately weak compared to the other forces. One possible explanation is that we experience only a fraction of the force of gravity because most of it acts in hidden extra dimensions.

The relentless pursuit of this fundamental particle was propelled further by particle physics. No particle can move with a speed faster than the speed of light in vacuum; however, there is no limit to the energy a particle can attain. In high-energy acceler-ators, particles normally travel very close to the speed of light. In these conditions, as the energy increases, the increase in speed is minimal. As an example, particles in the LHC move at 0.999,997,828 times the speed of light at injection (energy $= 450$ GeV) and 0.999999991 times the speed of light at top energy (energy $= 7000$ GeV). Therefore, particle physicists do not generally think about speed, but rather about a particle's energy. Energy and mass are two sides of the same coin. Mass can transform into energy and vice versa in accordance with Einstein's famous equation ($E = mc^2$), and because of this equivalence, mass and energy can be measured with the same unit (by setting $c = 1$). At the scale of particle physics these are the electronvolt and its multiples.

In this pursuit is where CERN became involved. Just because something looks like the Higgs particle does not mean it is the Higgs particle. If physicists do discover a new particle, they will need to measure its numerous properties before they can determine whether it is the Higgs boson described by the Standard Model of particle physics. Theory predicts in great detail how a Standard Model Higgs particle would interact with other particles. Only after carefully measuring and testing these interactions, like a biologist examining the genetic makeup of a new plant species would scientists be certain that they had indeed found the Standard Model Higgs boson. A new particle that did not act as expected would give physicists a whole new set of mysteries to explore.

The Standard Model is a collection of theories that embodies all of our current understanding of fundamental particles and forces. According to the theory, which is supported by a great deal of experimental evidence, quarks and leptons are the building blocks of matter, and forces act through carrier particles exchanged between the particles of matter. Forces also differ in their strength.

Standard Model of Elementary Particles

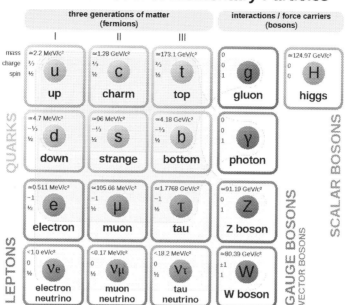

Quarks and Leptons are the building blocks which build up matter, i.e., they are seen as the "elementary particles" In the present standard model, there are six "flavors" of quarks. They can successfully account for all known mesons and baryons (over 200). The most familiar baryons are the proton and neutron, which are each constructed from up and down quarks. Quarks are observed to occur only in combinations of two quarks (mesons) and three quarks (baryons). There was a recent claim of observation of particles with five quarks (pentaquark), but further experimentation has not borne it out.

Quark	Symbol	Spin	Charge	Baryon number	S	C	B	T	Mass*
Up	U	1/2	+2/3	1/3	0	0	0	0	1.7−3.3 MeV
Down	D	1/2	−1/3	1/3	0	0	0	0	4.1−5.8 MeV
Charm	C	1/2	+2/3	1/3	0	+1	0	0	1270 MeV
Strange	S	1/2	−1/3	1/3	−1	0	0	0	101 MeV
Top	T	1/2	+2/3	1/3	0	0	0	+1	172 GeV
Bottom	B	1/2	−1/3	1/3	0	0	−1	0	4.19 GeV(MS)
									4.67 GeV(1S)

We are in the search of these quarks, protons, and electrons, which all have a low decay time from their creation, which is where the accelerator solution comes in.

Accelerators

A Large Hadron Collider (LHC) was built at CERN with the goal to smash protons moving at 99.999,999% of the speed of light into each other and so recreate conditions a fraction of a second after the big bang. The LHC experiments try and work out what happened in particle physics experiments. Particle physics is the unbelievable in pursuit of the unimaginable. To pinpoint the smallest fragments of the universe you have to build the biggest machine in the world. To recreate the first millionths of a second of creation you have to focus energy on an awesome scale. Some questions that we hope to answer with LHC experiments are as follows:

- Why do we observe matter and almost no antimatter if we believe there is a symmetry between the two in the universe?
- What is this "dark matter" that we cannot see that has visible gravitational effects in the cosmos?
- Why cannot the Standard Model predict a particle's mass?
- Are quarks and leptons actually fundamental, or made up of even more fundamental particles?
- Why are there exactly three generations of quarks and leptons?
- How does gravity fit into all of this?

To get the LHC experiments a set of outcomes including the discovery of the "God particle", we established the following frontiers:

- **The Energy Frontier**: using high-energy colliders to discover new particles and directly probe the architecture of the fundamental forces.
- **The Intensity Frontier**: using intense particle beams to uncover properties of neutrinos and observe rare processes that will tell us about new physics beyond the Standard Model.
- **The Cosmic Frontier**: using underground experiments and telescopes, both ground and space based, to reveal the natures of dark matter and dark energy and using high-energy particles from space to probe new phenomena.

The Large Hadron Collider (LHC) is located in a circular tunnel 27 km (17 miles) in circumference. The tunnel is buried around 100 m (about the size of a football field) underground. LHC straddles the Swiss and French borders on the outskirts of Geneva. The collider consists of distinct sets of components that cover the 27 km boundary. The goal of the collider is to deliver the smash of the two photons and provide introspections into the underlying matter and how fast it decays. Prior to the LHC we first built a **Large**

Electron–Positron Collider (**LEP**), one of the largest particle accelerators ever constructed.

LEP collided electrons with positrons at energies that reached 209 GeV. It was a circular collider with a circumference of 27 km built in a tunnel roughly 100 m (300 ft) underground and used from 1989 until 2000. Around 2001 it was dismantled to make way for the LHC, which reused the LEP tunnel. The LEP is the most powerful accelerator of leptons ever built. LEP was built with four detectors, each built around the four collision points within underground halls. Each was the size of a small house and was capable of registering the particles by their energy, momentum, and charge, thus allowing physicists to infer the particle reaction that had happened and the elementary particles involved. By performing statistical analysis of this data, knowledge about elementary particle physics is gained. The four detectors of LEP were called Aleph, Delphi, Opal, and L3. They were built differently to allow for complementary experiments. The colliders and detectors created and used were the following:

- **ALEPH** is an acronym for *Apparatus for **LEP PH**ysics at CERN*. ALEPH is a detector that determined the mass of the W-boson and Z-boson to within one part in a 1000. The number of families of particles with light neutrinos was determined to be 2.982 ± 0.013, which is consistent with the standard model value of 3.
- **DELPHI** Is an acronym for **DE**tector with **L**epton, **P**hoton, and **H**adron **I**dentification. Like the other three detectors, it recorded and analyzed the result of the collision between LEP's colliding particle beams. DELPHI was built in the shape of a cylinder over 10 m in length and diameter, and a weight of 3500 tons. In operation, electrons and positrons from the accelerator went through a pipe going through the center of the cylinder, and collided in the middle of the detector. The collision products then traveled outwards from the pipe and were analyzed by a large number of subdetectors designed to identify the nature and trajectories of the particles produced by the collision.
- **OPAL** is an acronym for *Omni-Purpose Apparatus for LEP*. The name of the experiment was a play, as some of the founding members of the scientific collaboration which first proposed the design had previously worked on the JADE detector at DESY in Germany. OPAL was designed and built as a general-purpose detector designed to collect a broad range of data. Its data were used to make high precision measurements of the Z-Boson lineshape, perform detailed tests of the Standard Model, and place limits on new physics.
- **L3** was another LEP experiment. Its enormous octagonal magnet return yoke remained in place in the cavern and became part of the ALICE detector for the LHC.

LEP was successful in many experiments and had provided enough directions to what physics directions it could base on the standard model, however we felt the need to create and build a larger collider which will give us beyond the standard model and potentially even help us move into the discovery of the elusive Higgs boson particle. The challenge is significant in the LEP that it has been built in a tunnel, reusing this tunnel will require us to decommission the LEP and then build the detectors and colliders for the LHC.

What the rebuild will provide us is an opportunity to align all technologies in the big data platform layers to collect data that is generated by these mammoth colliders, which will help us rebuild the collision process and visualize the impact repeatedly. The LEP collider was having Oracle databases, C programs and Open Source data models and platforms, Pascal, Fortran, and SAS but had to collect data as needed missing on foundational areas sometimes, and there were issues in the need for time as collisions occurred and people recorded as much as possible. To compute better with collection of large volumes of data and analyze results on demand the data platforms added at LHC included Hadoop, NoSQL, Spark, and Kubernetes. The next segment is to look at the technologies and the data.

Big data platform and application

History of LHC at CERN can be summarized in these events:

    1982: First studies for the LHC project
    1983: Z0/W discovered at SPS proton antiproton collider (SppbarS)
    1989: Start of LEP operation (Z/W boson factory)
    1994: Approval of the LHC by the CERN Council
    1996: Final decision to start the LHC construction
    2000: Last year of LEP operation above 100 GeV
    2002: LEP equipment removed
    2003: Start of LHC installation
    2005: Start of LHC hardware commissioning
    2008: Start of (short) beam commissioning
    2009: Repair, recommissioning, and beam commissioning
    2011: Launch of LHC

As LHC was being designed at CERN, the data calculations were being computed for each device and how much data is needed for use. CERN has approximately 3000 Members and 12,000 users. The Large Hadron Collider and four big Experiments: ATLAS, CMS, LHCb, and ALICE generate data $24 \times 7 \times 365$. CERN OpenLab is a public–private partnership, through which CERN collaborates with leading ICT companies and research organizations. The Worldwide LHC Computing Grid (WLCG) is a global collaboration of more than 170 institutions in 42 countries which provide resources to store, distribute, and analyze the multiple PBs of LHC Data. Data at CERN 15 PB per month and >250 PB today at CERN data center.

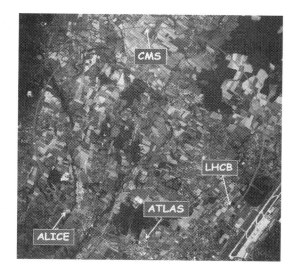

CERN Implementation of LHS

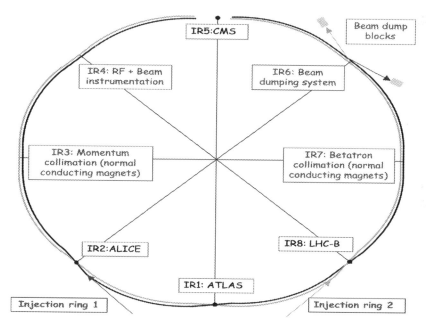

Data Generation at LHC

The picture shown above is the data generation at LHC. This data is generated for each execution and comes in multiple formats including the following: ROOT for Physics, Parquet, JSON, TSV, CSV, Log files, and XML formats. We need to collect the data for analytics executed on HDFS on demand, we also need to execute stored compute on database platforms and we need to execute streaming analytics in memory as data streams. The challenge here is that we will collect several terabytes of data from source generated files, but need to provide 100 −200 GB new extracts for analytics, while we will still have access to operational data for running analytics and exploration.

To process data the new platforms to add included Apache Hadoop, Apache Kafka, Apache Spark, Apache Flume, Apache Impala, Oracle, and NoSQL database. This data processing architecture will be integrated with the existing ecosystem of Oracle databases, SAS, and Analytics systems. The Apache stack selected is shown in the picture below.

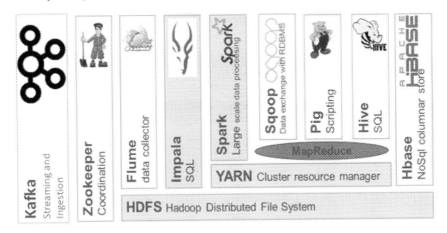

Hadoop configuration implemented at CERN includes the following:

- Baer Metal Hadoop/YARN Clusters
  - five Clusters
  - 110 + nodes
  - 14 + PBs Storage
  - 20 + TB Memory
  - 3100 + Cores
  - HDDs and SDDs

Access to data is provided with Active Directory and native security rules are enclosed for each layer of the access from the Grid to Hadoop. The rules provide encryption, decryption, hierarchies, and granularity of access. The authorization policy is implemented in the rules and the authentication is implemented as Active Directory.

The end user analysts and physicists at CERN use Jupyter notebooks with PySpark implementation to work on all the data. The Jupyter notebooks use Impala, Pig, and Python and several innovations have been added by the CERN team to use the Apache stack for their specific requirements. We will discuss these innovations in the next segment.

Innovations:

XRootD filesystem interface project

The CERN team evaluated the Apache stack and identified a few gaps between where they were with current technology and the new stack to be augmented. The gaps were all physics files were written using the ROOT project and this project was developed in c++ and formats will not be able to load into AVRO or Spark. The CERN team joined hands with DIANA-HEP team to create the XRootD project. The project was designed to load physics files into HDFS and Spark. Details of the project can be found at http://xrootd. org and the GitHub page for the project is at https://github.com/cerndb/hadoop-xrootd.

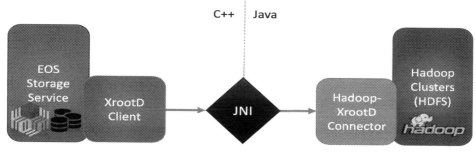

XRootD Project

XRootD: The XRootD project aims at giving high performance, scalable fault tolerant access to data repositories of different kinds, and the access will be delivered as file based. The project was conceived to be delivered on a scalable architecture, a communication protocol, and a set of plug-ins and tools based on those. The freedom to configure XRootD and to make it scale (for size and performance) allows the deployment of data access clusters of virtually any size, which can include sophisticated features, like authentication/authorization, integrations with other systems, and distributed data distribution. XRootD software framework is a fully generic suite for fast, low latency, and scalable data access, which can serve natively any kind of data, organized as a hierarchical filesystem-like namespace, based on the concept of directory.

Service for web-based analysis (SWAN)

CERN has packaged and built a service layer for analysis based on the web browser. This service called SWAN is a combination of the Jupyter notebook, Python, C++, ROOT, Java, Spark, and several other API interfaces. The package is available for download and usage for any consumer who works with CERN. The SWAN service is available at https:// swan.web.cern.ch.

There are several other innovations to manage the large files, the streaming analytics, the in-memory analytics, and kerberos security plug-ins.

The result—Higgs Boson discovery

The discovery of the Higgs particle in 2012 is an astonishing triumph of mathematics' power to reveal the workings of the universe. It is a story that has been recapitulated in physics numerous times and each new example thrills just the same. The possibility of black holes emerged from the mathematical analyses of German physicist Karl Schwarzchild; subsequent observations proved that black holes are real. Big Bang cosmology emerged from the mathematical analyses of Alexander Friedmann and also Georges Lemaître; subsequent observations proved this insight correct as well. The concept of antimatter first emerged from the mathematical analyses of quantum physicist Paul Dirac; subsequent experiments showed that this idea, too, is right. These examples give a feel for what the great mathematical physicist Eugene Wigner meant when he spoke of the "unreasonable effectiveness of mathematics in describing the physical universe." The Higgs field emerged from mathematical studies seeking a mechanism to endow particles with mass, and once again the math has come through with flying colors.

Nearly a half-century ago, Peter Higgs and a handful of other physicists were trying to understand the origin of a basic physical feature: mass. You can think of mass as an object's heft or precisely as the resistance it offers to having its motion changed. Accelerate a car to increase its speed, and the resistance you feel reflects its mass. At a microscopic level, the car's mass comes from its constituent molecules and atoms, which are themselves built from fundamental particles, electrons, and quarks. But where do the masses of these and other fundamental particles come from?

When physicists in the 1960s modeled the behavior of these particles using equations rooted in quantum physics, they encountered a puzzle. If they imagined that the particles were all massless, then each term in the equations clicked into a perfectly symmetric pattern, like the tips of a perfect snowflake. And this symmetry was not just mathematically elegant. It explained patterns evident in the experimental data. But here is the puzzle, physicists knew that the particles did have mass, and when they modified the equations to account for this fact, the mathematical harmony was spoiled. The equations became complex and unwieldy and, worse still, inconsistent. What to do?

Here is the idea put forward by Higgs. Do not shove the particles' masses down the throat of the beautiful equations. Instead, keep the equations pristine and symmetric, but consider them operating within a peculiar environment. Imagine that all of space is uniformly filled with an invisible substance, now called the Higgs field that exerts a drag force on particles when they accelerate through it. Push on a fundamental particle in an effort to increase its speed and, according to Higgs, you would feel this drag force as a resistance. Justifiably, you would interpret the resistance as the particle's mass. For a mental toehold, think of a ping pong ball submerged in water. When you push on the ping pong ball, it will feel much more massive than it does outside of water. Its interaction with the watery environment has the effect of endowing it with mass. The same is the case of explanation with particles submerged in the Higgs field.

The physics community had, for the most part, fully bought into the idea that there was a Higgs field permeating space. Mathematical equations can sometimes tell such a convincing tale; they can seemingly radiate reality so strongly, that they become entrenched in the vernacular of working physicists, even before there is data to confirm them. But it is only with data that a link to reality can be forged. How can we test for the Higgs field?

This is where the Large Hadron Collider (LHC) comes in. Winding its way hundreds of yards under Geneva, Switzerland, crossing the French border and back again, the LHC is a nearly 17-mile-long circular tunnel that serves as a racetrack for smashing together particles of matter. The LHC is surrounded by about 9000 superconducting magnets, and is home to streaming hordes of protons, cycling around the tunnel in both directions, which the magnets accelerate to just shy of the speed of light. At such speeds, the protons whip around the tunnel about 11,000 times each second and when directed by the magnets, engage in millions of collisions in the blink of an eye. The collisions, in turn, produce fireworks-like sprays of particles, which mammoth detectors capture and record.

One of the main motivations for the LHC, which cost on the order of $10 billion and involves thousands of scientists from dozens of countries, was to search for evidence for the Higgs field. The math showed that if the idea is right, if we are really immersed in an ocean of Higgs field, then the violent particle collisions should be able to jiggle the field, much as two colliding submarines would jiggle the water around them. And every so often, the jiggling should be just right to flick off a speck of the field a tiny droplet of the Higgs ocean which would appear as the long-sought Higgs particle.

The calculations also showed that the Higgs particle would be unstable, disintegrating into other particles in a minuscule fraction of a second. Within the maelstrom of colliding particles and billowing clouds of particulate debris, scientists armed with powerful computers would search for the Higgs' fingerprint, a pattern of decay products dictated by the equations.

In the early morning hours of July 4, 2012, as the world came to quickly learn, the evidence that the Higgs particle had been detected was strong enough to cross the threshold of discovery. With the Higgs particle now officially found, physicists worldwide broke out into wild applause. Peter Higgs wiped away a tear.

The Higgs particle represents a new form of matter, which had been widely anticipated for decades but had never been seen. Early in the 20th century, physicists realized that particles, in addition to their mass and electric charge, have a third defining feature: their spin. But unlike a child's top, a particle's spin is an intrinsic feature that does not change; it doesn't speed up or slow down over time. Electrons and quarks all have the same spin value, while the spin of photons, particles of light is twice that of electrons and quarks. The equations describing the Higgs particle showed that unlike any other fundamental particle species it should have no spin at all. Data from the Large Hadron Collider have now confirmed this.

Establishing the existence of a new form of matter is a rare achievement, but the result has resonance in another field: cosmology, the scientific study of how the entire universe began and developed into the form we now witness. For many years, cosmologists studying the Big Bang theory were stymied. They had pieced together a robust description of how the universe evolved from a split second after the beginning, but they were unable to give any insight into what drove space to start expanding in the first place. What force could have exerted such a powerful outward push? For all its success, the Big Bang theory left out the bang. The LHC's confirmation that at least one such field actually exists thus puts a generation of cosmological theorizing on a far firmer foundation.

Lessons Learned: The significant set of lessons we have learned in discussing the CERN situation and its outcomes with Big Data Analytics implementation and the future goals include the following:

Problem Statement: Define the problem clearly, including the symptoms, situations, issues, risks, and anticipated resolutions. The CERN team started this process since the inception of the LEP and throughout the lifecycle of all its associated devices; they also defined the gaps and areas of improvement to be accomplished which were all defined in the LHC process.

Define solution: this segment should identify all possible solutions for each area of the problem. The solution segment can consist of multiple tools and heterogenous technology stacks integrated for a definitive, scalable, flexible, and secure outcome. The definition of the solution should include analytics, formulas, data quality, data cleansing, transformation, rules, exceptions, and workarounds. These steps will need to be executed for each area and include all the processes to be defined in clarity. CERN team has implemented this and added governance to ensure that the steps are completed in accordance and no gaps are left unanswered, and if gaps exist there are tags and tasks associated with the tags for potential completion.

Step by step execution: is a very essential mechanism to learn how to become successful. If you read the discovery of the Higgs field, the experiment proves that we need to iterate multiple times for every step to analyze the foundational aspects, which will provide us more insights to drill through to greater depths. This step by step process is very much seen to bring success, whether we work on cancer research or in-depth particle physics research the concept to proof perspective demands steps and outcomes at each step, adjustments to be made recorded and the step reprocessed and outcomes recorded.

In big data applications the step by step execution is very much possible with the data collected in HDFS at the raw operational level, which can be explored, discovered, experimented and constructed in multiple methods, cycles and analysis of the details performed for the data. All of these are possible within the HDFS layers which provides us the playground to prove the possibilities. The cost models are not necessarily cheap, CERN for example has spent over $1B on infrastructure worldwide over the years, but one Higgs field result is more valuable than the spend, and the recovery is multifold. Here is where the technology has and is enabling the experiments to head towards:

- 2010−12: Run 1 ;7 and 8 TeV
- *2015−1 : Run 2 ; 13 TeV*
- 2021−23: Run 3 (14 TeV)
- 2024−25: HL-LHC installation
- 2026−35: HL-LHC operation

The next iteration of this big data application is the introduction of cloud and kubernates platforms where more experiments can be conducted at each component of the LHC.

Open source adoption: has been a critical success factor in the CERN application of Big Data platforms and applications. The open source licenses and security have evolved to be used for enterprise applications and can be deployed on a global basis today for which CERN is one proof. This is a critical success factor to remember.

Governance: is the other key critical success factor to this entire journey and anybody in the scientific community will agree to this viewpoint. Governance is not about being measured but reporting on the measurements required for ensuring all activities and their tasks have been completed, per the expected outcomes and times completed. This governance enabled CERN to become successful from the first iteration to date and will be used in the future.

5

Pharmacy industry applications and usage

Torture the data, and it will confess to anything.
Ronald Coase, winner of the Nobel Prize in Economics

Pharmaceuticals run extremely complex mathematical analytics compute across all their processes. The interesting viewpoint here is the fact that they are dwelling in a world of data complexity, which needs to be understood across different layers with the appropriate blending of insights. The incorrect application of formulas and calculations will only misguide us to incorrect conclusions. Data is something that you have to manipulate to get at key truths. How you decide to treat your data can vastly affect what conclusions you make.

In the use case analysis in this chapter, we will discuss the implementation of Hadoop by Novartis, and their approach to overcome challenges faced in the traditional data worlds for complexity of compute and multi-user access of the underlying data across the same time for other analytics and reporting. We will discuss the facets of big data applications looking into accessing streaming data sets, data computations using in-memory architectures, distributed data processing for creating data lakes and analytical hubs, in-process visualizations and decision support, the data science team and how the change needs to happen for successful creation of big data applications. We will discuss the usage and compliance requirements for the data, the security, encryption, storage, compression, and retention specific topics as related to pharmaceutical industry.

Complexity is a very sensitive subject in the world of data and analytics. By definition it deals with processes that are interconnected and have dependencies that may be visible or hidden, often leading to chaos in the processing of the data. These systems have exhibited characteristics including but not limited to the following:

- The number of parts (and types of parts) in the system and the number of relations between the parts is nontrivial. There is no general rule to separate "trivial" from "nontrivial", and it depends on the owner of the data to define these rules and document them with flows and dependencies. This issue is essential to deal with as large systems can get complex and become less used, which is both cost and productivity loss.
- The system has memory or includes feedback and loops which are needed to be defined and exit strategies need to be validated for each operation. Scientific compute falls into this category and there are several case studies of experiments

Building Big Data Applications. https://doi.org/10.1016/B978-0-12-815746-6.00005-3

and discoveries that need to be revalidated due to lack of clarity on the process complexity.

- The system can adapt itself according to its history or feedback. This is another vital complexity that needs to be handled, and the system self-adaptability needs to be defined and documented as requirements and the specific outcomes for this issue needs to be articulated with fine grain details, which will help the process complexity be handled. These aspects are not handled with the greatest set of requirements processes and often lead to delays of systems being accepted during the user acceptance testing stages, whether we do waterfall or agile development.

- The relations between the system and its environment are nontrivial or nonlinear. This aspect is very important to understand as the internet of things and big data applications today provide data from anywhere all the time. There is a need to understand that changes can occur in the data in the nonlinear relationship, which needs to be ingested and processed, linked to the prior state and analytics need to be computed based on the change of data, this complexity is to be defined and processed. Whether the change is trivial or nontrivial needs to be defined by the aspect of data we are working with, for example if we are monitoring the stock markets, and we get streaming feeds to fake and real information, who can validate the data and what aspects need to be put on an alert? If we do go ahead with the data and it turns out to be fake information, who can validate and clarify? This simple process is complexity for data in the world we live today in and the internet makes it all the more complex as it provides the news $24 \times 7 \times 365$ to the world.

- The system is highly sensitive to initial conditions. In scientific research this complexity is very much present every minute. The reason why it is complex is because the initiator of any experiment has a specific set of conditions that they want to study and collect behaviors of the system. This data set is very relevant to the person studying the data, but the initial sensitivity needs to be documented to provide relevance to other users of the data or the system. The initial conditions are essential in scientific research and numerous applications of this data with big data platforms will allow the application to be built following the steps of each stage and this data can be captured in specific files and reused to understand the outcomes. We did not have this capability in the traditional data world due to the formats and compliance of data structures.

The data and the system complexity are further understood as we look at the layers of data that we will need to assimilate for building applications. The data layers here will need to be defined, managed, transformed, logged, aggregated, and be available. Let us see this series, and this is described in the figure below (Fig. 5.1).

As seen in the picture above, we have the innate capability to take a data at its raw granularity layer as a small particle and by the time it becomes a layer in the analytics, we apply transformations and aggregations, decide to change formats, leave attributes

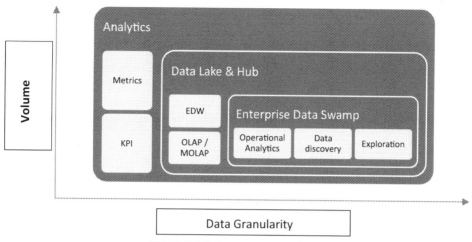

FIGURE 5.1 Volume versus granularity.

out as they might not be significant in the layer needed, resulting in complexity of lineage, and often nonadoption of the data and its associated insights.

Formulation of the data across the layers of compute is very intricately complex and in the traditional world of ETL we often have to add trace to detect if there are rejects or extremely complex formula of calculations causing both performance and traceability issues. We tend to either ignore the issue or it is added to a list of bug fixes, which never happen or happen over a long period of time. This issue of complexity becomes a nightmare when we deal with analytics.

Analytics in any enterprise whether small or large is an essential topic. The core foundations of the enterprise are depending on its analytics, whether we talk about earnings, losses, expenses, incomes, or stock prices. We measure and depending on the measure outcomes, we work with the data and decide the performance of the enterprise. In order to ensure sustained positive measures, the enterprise measures its customers' journeys, its products' journeys, its supply chain journeys, its research journeys, its competitive journeys, its marketing journeys, its sales journeys, and even its operational journeys. We have become obsessed with the measurements and metrics, that we even measure the time it takes to place an order or receive a pick-up. These metrics and measure aspects increase even more in the world of internet of things, the research and innovation areas, healthcare and medicine-related research, patient treatment–related research and forays into space and universes are all magnitudes of complexity to comprehend and deliver. Analytics can provide a tipping point, but to get to that tipping point, we need to figure out the methods and techniques. This is where we will introduce you to a new method of defining and designing for complexity. It is a part of the new data management strategy and architecture process.

The complexity design for data applications

A new methodology can be incorporated for understanding complexity and designing the architecture to handle the same. The methodology is a combination of steps migrated from traditional techniques of data processing along with new steps. Here is a view of the methodology.

The steps for managing complexity are shown in Fig. 5.2. These steps can be phased and sequenced depending on the infrastructure and the type of data we need to work with. In the world of big data, the complexity of the data is associated with its formats and the speed of the production of data. Let us start with understanding the sequencing of these steps; the data is streaming data from either an airplane or a critical patient, which comes with several metadata elements embedded in the data as it originates from the source. We need to understand how to deal with this data, learn from the discoveries the aspects that it portrays for use in compute and insight delivery.

The first step in this process to understand the myriad complexity with the data we receive. To understand the data there are two ways, one is to collect sample data sets and understand the data, the second is to stream the data and setup algorithms to understand the data. The second technique is what big data algorithms will help us facilitate. The big data algorithms that manage and process streaming data are designed and implemented to process, tag, and analyze large and constantly moving volumes of data. There are many different ways to accomplish this, each with its own advantages and disadvantages. One approach is the native stream processing also called as tuple-at-a-time processing. In this technique, every event is processed as it comes in, one after the other, resulting in the lowest-possible latency. Unfortunately, processing every incoming event is also computationally expensive, especially if we are doing this in-memory and the data is complex by nature. The other technique is called the micro-batch processing, and in this approach, we make the opposite tradeoff, dividing

Data Acquisition
Data Discovery
Data Exploration
Data Attribution & Definition
Data Analysis
Data Tagging
Data Segmentation & Classification
Data Storyboarding
Data Lake Design
Data Hub DEsign
Analytics

FIGURE 5.2 Data complexity methodology.

incoming events up into small batches either by arrival time or until a batch has reached a certain size. This reduces the computational cost of processing but also introduces more latency. To determine the use of streaming data, ask if the value of the data decreases over time. If the answer is yes, then you might be looking at a case where real-time processing and analysis is beneficial.

An exciting application of streaming data is in the field of machine learning, which is a hot subject area across big data platforms and cloud computing today. One of the benefits of streaming data is the ability to retrain the algorithms that are used in machine learning, as unsupervised learning reinforces the algorithm to learn as new data becomes available in streams. These patterns and identities of data that are collected and stored as data processing occurs will be used in the data attribution and definition process. The data collected will also contain associated metadata which will be useful in defining the architecture of the table and the associated file in the data lake and further application areas.

Another approach to understand the complexity of the data is to execute a data discovery process. In this step, the data is acquired and stored as raw data sets. The acquisition process if not streaming, can use Apache Kafka or Apache NiFi processes to define the structure and layout of the data. Understanding this is essential in data discovery as we will run through several iterations of analysis of the data received, the time of its receipt, the business value of the data, the insights that can be delivered. Data discovery is an interesting exercise as you need to know what you are looking for. This technique is useful in reading log files, process outcomes, and decision support mechanisms. In research the technique is excellent to identify the steps and outcomes for any experiments. The benefit of this approach is the ability for the end user to deliver the complexity of the data and its usefulness. We will do data discovery even if we execute streaming data analysis, this is a must do step for realizing the overall structure, attribution, formats, value and insights possible from the data acquired.

Another key step once we execute data discovery process is the data exploration. In this process we will look through the data and see where all will any connects of the data with other data occur. The exploration of data is essential to determine not only the connect possibilities, but also to see if the data has issues including but not limited to the following:

- Duplicate information
- Missing data
- Random variables
- Format issues
- Incorrect rounding or float termination

These issues can be discovered during data exploration and be identified for evaluation and source systems can be notified if these issues need to be fixed prior to next feed of data from the same system. An interesting example is the usage of drugs for treatment of patients; let us assume the same Benadryl is being given to pediatric patients and adult patients. The system feeds for patient vitals and drugs administered are electronic, and in this situation if the dosage is not identified as pediatric or adult, it will be confusing on how to align outcomes and readings, if the actual dosage information was not captured or was missing or was incorrectly stated. This type of a situation calls for earlier data discovery or exploration and that is just what the big data platform provides to us. We can use interactive query applications like Tableau or Apache Drill or Apache Presto to run the exercises. The preferred format of data is JSON, which provides us the flexibility to add more values as needed or correct the values as data is ingested and explored.

Once we have completed data discovery and data exploration, we are now at a stage where we can attribute and define the data. In the world of applications, data attribution and definition need to be done from the raw data layers and through each stage of transformation that data needs to evolve. The attribution of data includes the identification of the data, its metadata, its format, and all values associated with the data in the samples or the actual data received. The attribution process will be important to structure the new data which can be from any source and have any format. The attribution will deliver the metadata and define the formats.

The next step to identify and define complexity is completed by the analysis of data after the attribution process. The complexity of data is identified by the analysis of the data. What complexity are we talking about is the incompleteness of data and missing data that needs to be defined in the analysis. The dependency on those aspects of data is very important to understand from both the analytics and the application usage perspectives. This complexity haunts even the best designed data warehouse and analytics platforms. In the big data application world, the underlying infrastructure provides us the opportunity to examine the raw data and fix these issues. This complexity needs to be managed and documented.

Another complexity that needs to be defined and managed is the source data from multiple sources colliding and causing issues. The business teams' rules are defined once the data has been defined and the issue needs to be sorted out, but what about the raw data? In the world of big data applications this issue needs to be handled at the raw data level too as data discovery and exploration, and operational analytics are all executed in the raw data layer. To manage this complexity, we will be tagging each source row, file, or document. The rows or files or documents that are conflicting will be available for inspection and the users can determine what they want to do with the data and recommend the rules for ingestion of the data.

The next step to manage complexity is to classify and segment the data. This process will classify the data by the type of data, the format and its complexity, and segment the data, files and associated documents into separate directories for management of the data from the raw data layer to be managed across till the application layer.

Complexities in transformation of data

There are several states of transformation we will make on the data from the time it is available till it is used in an application. These transformations will carry several integrations and business rules applied to the data, and often we lose sight of the number of transformations and their hidden formulas and associated rules to integrate the data. This complexity needs to be very clearly defined and documented, especially in the new world of data where we will be integrating the machine learning, artificial intelligence, and several neural network algorithms, analytics, and models. While we can learn and fix issues as we come across them, the data volume is different, the formats are vast and different, the infrastructure is resilient and can scale out as much needed, yet the complexity needs to be defined for both the understanding of what is being done and the associated outcomes.

The transformations will occur at different layers of the data architecture; we will run transformations for operational analytics, data discovery and data exploration at the raw data swamp layer. These transformations will be executed by multiple teams and multiple end users, and outcomes from these exercises will be validated for use of the data. In this case the complexity is validated and will be documented if the data use case is accepted for further analysis.

Transformations will occur further as we start the journey to data lake. This is an enterprise asset and will be used by all users for executing the business reports, extracting insights, and delivering more integration touchpoints. The transformations here will be implemented as microservice architecture, which means we need to define and design complexity within the libraries used in the microservices layers. The same transformation exercises and integration exercises will occur in the data hub and it will require the complexity to be broken down to manageable pieces of architecture. The final layer of transformations is the analytical modules where we will use artificial intelligence, machine learning, neural network algorithms, and analytical models. These layers deliver fantastic results but need the appropriate inputs to be applied with the right granularity and data quality. This is another layer to manage complexity once the data is available for compute.

Now that all possible layers and associated discussions have been had on complexity, let us see the real-life management of data in the pharmaceutical industry.

There are several distinct use cases in the pharmaceutical industry that we will discuss, these include the following:

- Drug discovery
- Patient clinical trials
- Social media community
- Compliance

Drug discovery is a very intricate and complex process, which needs the researchers to develop a comprehensive understanding of how the human body works at the molecular level. This means to develop a thorough understanding of how the body reacts to current treatments, document the in-process experiments on the different studies of changes being developed to drugs, and have a much better grasp of the killer-effects from consumptions of drugs including side effects and intricacies caused. All of this can be made feasible with seamless collaboration between the industry, academia, the regulators, governments, and healthcare providers. But the issue here is there is too much data floating around, not all of it is used, we still struggle with integration of technology to answer the foundational problems. There is a solution with big data platforms, and we have seen several leading pharmaceuticals unraveling the discoveries made with the use of big data infrastructure.

To understand how the big data platforms become useful, we need to take a peek into history. In the early 1990s there was a definition called as Eroom's law, which primarily is the observation that drug discovery was becoming slower and more expensive over time. This slowdown was happening despite improvements in technology including high throughput screening, biotechnology and chemistry, and computational drug-design. The cost of developing a new drug roughly doubles every 9 years with all inflation adjustments was the definition outcome from the observation. The name of the law was the reverse of Moore's law in technology, to show the issue. There are significant areas that led to the observation being formed and these included the following:

- **Minimal incremental effect**: an opinion that new drugs only have modest incremental benefit over drugs already widely considered as successful, and treatment effects on top of already effective treatments are smaller than treatment effects versus placebo. The smaller size of these treatment effects mandates an increase in clinical trial sizes to show the same level of efficacy, which results in longer delays and inefficiencies that were not integrable until big data application, mashups, and platforms became a reality.
- **A cautious regulator issue**: the progressive lowering of risk tolerance seen by drug regulatory agencies makes drug research and discovery both costlier and harder. After older drugs are removed from the market due to safety reasons, the bar on safety for new drugs is increased, which makes it more expensive for the process to complete and this causes a slowdown.
- **Spend more and get where?** the tendency to add human resources and system resources to R&D, which may lead to project overrun. This is a very precarious situation where you are at a point of no return. This is eliminated in the new landscape to a large extent.
- **The brute force bias**: the tendency to overestimate the ability of advances in basic research and brute force screening methods to show a molecule as safe and effective in clinical trials. From the 1960s to the 1990s (and later), drug discovery has shifted from whole-animal pharmacology testing methods to reverse pharmacology target-approaches that result in the discovery of drugs that may tightly bind with high-affinity to target proteins, but which still often fail in clinical trials due to an underappreciation of the complexity of the whole organism. This issue is a brute force technique which provides limited success if at all and often has led to more failure.

How big data applications change the landscape: a very interesting situation to look into is the big data landscape changing the entire drug discovery process. Take for example the application to visualize the whole-body protein targets for a given disease, using a specific combination of drugs. This application will require us to mashup different areas of data, from discovery to clinical trials to actual outcomes. The mashup will require us to use an application layer like Tableau, with Apache Drill as an interconnection to federate the data from different source platforms or subject areas. The subject areas can be all in one system across different directories or different systems, the query integration will execute each query independently and combine the outcomes as needed for the mashup. This gives rise to a fantastic idea, where we realize that by integrating a specific set of subject areas with a key focus target set, we can easily create a drug discovery platform with outcomes defined and managed using technology. This disruption is what we have been waiting for a number of years, and the disruption is resilient in its incorporation into the enterprise. We have read multiple case studies from Novartis, GlaxoSmithKline, Johnson&Johnson, Merck, and others on their successful transformations.

The key focus areas shown in Fig. 5.3, which we are discussing includes the following:

- Current drug and composition
- New drugs and composition
- Patient recruitment
- Patient adherence
- Segmentation of diagnosed and undiagnosed patients
- Integration of social media data
- Improve safety signal detection
- Disease pattern analysis
- Flexible and faster design of products, trials, and treatments

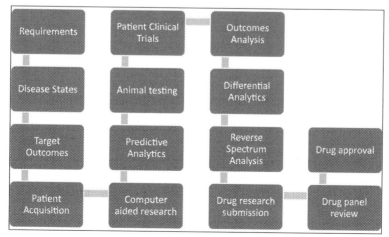

FIGURE 5.3 Drug discovery and research.

- Minimal animal testing and dependency
- Refine the trials list to the most viable targets
- Real world data (RWD)
- Deploy in cloud
- Design patient diary for tracking and compliance
- Electronic medical record (EMR)

By creating a series of data files with JSON or XML, we will bring data from these subject areas into the big data infrastructure. Our next goals are to create a series of predictive models and outcomes which will plot the expected results from each experiment and outcome. This model will accelerate the overall sharing and collaboration of data and experiments which will be used by multiple teams and their associated experiments. This type of data collaboration will provide us a pace of innovation that benefits us as a community. How did we get this far? The ideation is to have a file system–based data platform that can be leveraged to store large and small data files, all these files will contain data from drugs, patients, diseases, experiments, predictive outcomes, actual outcomes, expected end results, community predictions, community anticipations, community classification of disease stages, community response, community social media outreach, government agencies inspection, compliance and regulatory filings, actual patient reports and filings, actual providers research data, and actual overall outcomes. All of these mean that we still do drug discovery, but the process optimizes and transforms like this.

The process model is followed based on specific research and drug outcomes as desired by the pharmaceutical. The goal here is to beat the disease state and discover cure that can help the management of the disease. We have progressed with more drug discovery in the last few years based on genomic treatment with a patient specific state of treatment based on their gene and the reaction to the drug. This work is done once the gene is extracted from the patient and then research is driven in the lab, and outcomes are studied based on several factors, with the underlying behavior driven by the protein, RNA, DNA, and other molecular structures.

The continued success we are seeing today is aided with machine learning, neural networks, artificial intelligence, molecular research, bioscience, statistics, and continues to improve and generate global collaboration.

In all of these processes discussed so far, now bring in the complexity management and look at the processes and steps described. We can easily implement the management of complexity from acquisition of data, through ingestion, integration, segmentation, classification, analysis, operational analytics, predictive analytics, data lake transformations, data hub transformations, analytics, and visualizations. Most of the industries in pharmaceutical area today have a combination of heterogenous technologies and they all have made vast improvements in data strategy and management. With the evolution of cloud computing and microservices architecture, we are seeing more improvements in the reduction of complexity.

The next segment of pharmaceutical industry usage of big data applications and platform is in the integration of social media data, from different communities of patients who are working with them on several clinical trials and the experiences of the patients with the therapies at home and nonhospital surroundings The application layer integrates all the social media data, and it acquires the data with several tags that will link it to the specific patient, and align their responses in the community portals to their records. The patient data is aligned with all their sentiments, outcomes, recordings, experiences, and overall responses. Once this data is loaded, there are several artificial intelligence algorithms that work on interpreting the analytical outcomes and provide several prediction and prescription models. The application layer uses the mashup and leverages the data integration in the underlying infrastructure to create the magical outcome. The social media integration has accelerated drug clinical trials and reduced cycles from beyond 20 years in multiple phases to agile cycles with outcomes in less than 10 years at a maximum. It used to take forever to finish clinical trials, before the initial release, and this has been changed for good.

The focus of pharmaceutical industry has shifted to patient and has transformed to patient reactions and outcomes. The only way this focus has transformed is due to several large research teams from IBM, Google, and other industries contributing to several technologies, algorithms, neural networks, and more.

Google deep mind

Each scan and test result contain crucial information about whether a patient is at risk of a serious condition, and what needs to be done. Being able to interpret that information quickly and accurately is essential for hospitals to be able to save lives. AI systems could be hugely beneficial in helping with this process. Rather than programming systems by hand to recognize the potential signs of illness, which is often impossible given the number of different factors at play, AI systems can be trained to learn how to interpret test results for themselves. In time, they should also be able to learn which types of treatments are most effective for individual patients.

This could enable a series of benefits across healthcare systems, including the following:

1. **Improved equality of access to care:** Demands on these healthcare systems are felt more acutely in certain areas of the world, and even within certain departments in hospitals, than others. If we can train and use AI systems to provide world-class diagnostic support, it should help provide more consistently excellent care.
2. **Increased speed of care:** We hope that AI technologies will provide quick initial assessments of a patient to help clinicians priorities better, meaning patients go from test to treatment faster.
3. **Potential for new methods of diagnosis:** AI has the potential to find new ways to diagnose conditions, by uncovering and interpreting subtle relationships between

different symptoms and test results. In theory, this could lead to even earlier diagnosis of complex conditions.
4. **Continual learning and improvement:** Because AI tools get better over time, they will also help hospitals to continually learn about the approaches that help patients most.

Case study

- Novartis Institutes for Biomedical Research (NIBR), the global pharmaceutical research organization for Novartis. NIBR takes a unique approach to pharmaceutical research—at the earliest stages, analyzing and understanding the patient need, and disease specifics and responses, which align and help them determine their research priorities. On any given day, their scientists are working hard at nine research institutes around the world to bring innovative medicines to patients. Over 6000 scientists, physicians, and business professionals work in this open, entrepreneurial, and innovative culture that encourages true collaboration. One of NIBR's many interesting drug research areas is in Next Generation Sequencing (NGS) research. NGS research requires a lot of interaction with diverse data from external organizations such as clinical, phenotypical, experimental, and other associated data. Integrating all of these heterogeneous datasets is very labor intensive, so they only want to do it once.

One of the challenges they face is that as the cost of sequencing continues to drop exponentially, the amount of data that's being produced increases. Because of this, Novartis needed a highly flexible big data infrastructure so that the latest analytical tools, techniques, and databases could be swapped into their platform with minimal effort as NGS technologies and scientific requirements change. The Novartis team chose Apache Hadoop platform for investigating and discovery of data and its associated relationships and complexities.

NGS data requires high data volumes that are ideal for Hadoop, a common problem is that researchers rely on many tools that don't work on native HDFS. Since these researchers previously couldn't use systems like Hadoop, they have had to maintain complicated "bookkeeping" logic to parallelize for optimum efficiency on traditional High-Performance Computing (HPC). This workflow system uses Hadoop for its performance and robustness and to provide the POSIX file access (MapR Hadoop) that lets bioinformaticians use their familiar tools. Additionally, it uses the researchers' own metadata to allow them to write complex workflows that blend the best aspects of Hadoop and traditional HPC. The team then uses Apache Spark to integrate the highly diverse datasets. Their unique approach to dealing with heterogeneity was to represent the data as a vast knowledge graph (currently trillions of edges) that is stored in HDFS and manipulated with custom Spark code.

This innovative use of a knowledge graph lets Novartis bioinformaticians easily model the complex and changing ways that biological datasets connect to one another, while the use of Spark allows them to perform graph manipulations reliably and at scale.

On the analytics side, researchers can access data directly through a Spark API, or through a number of endpoint databases with schemas tailored to their specific analytic needs. Their tool chain allows entire schemas with 100 billion of rows to be created quickly from the knowledge graph and then imported into the analyst's favorite database technologies. As a result of their efforts, this flexible workflow tool is now being used for a variety of different projects across Novartis, including video analysis, proteomics, and metagenomics.

A wonderful side benefit is that the integration of data science infrastructure into pipelines built partly from legacy bioinformatics tools can be achieved in mere days, rather than months. By combining Spark and Hadoop-based workflow and integration layers, Novartis' life science researchers are able to take advantage of the tens of thousands of experiments that public organizations have conducted, which gives them a significant competitive advantage.

Additional reading

Predict malaria outbreaks (http://ijarcet.org/wp-content/uploads/IJARCET-VOL-4-ISSUE-12-4415-4419.pdf).

MIT Clinical Machine Learning Group is spearheading the development of next-generation intelligent electronic health records (http://clinicalml.org/research.html).

MATLAB's ML handwriting recognition (https://www.mathworks.com/products/demos/machine-learning/handwriting_recognition/handwriting_recognition.ht technologies).

Google's Cloud Vision API (https://cloud.google.com/vision/) for optical character recognition.

Stat News (https://www.statnews.com/2016/10/03/machine- learning-medicine-health/).

Advanced predictive analytics in identifying candidates for clinical trials (http://www.mckinsey.com/industries/pharmaceuticals- and-medical-products/our-insights/how-big-data-can- revolutionize-pharmaceutical-r-and-d).

The UK's Royal Society also notes that ML in bio- manufacturing for pharmaceuticals (http://blogs.royalsociety.org/in- verba/2016/10/05/machine-learning-in-the- pharmaceutical-industry/).

Microsoft's Project Hanover (http://hanover.azurewebsites.net/) is using ML technologies in multiple initiatives, including a collaboration with the Knight Cancer Institute (http://www.ohsu.edu/xd/health/services/cancer/) to develop AI technology for cancer precision treatment).

6

Visualization, storyboarding and applications

Most people use statistics the way a drunkard uses a lamp post, more for support than illumination
Mark Twain

Visualization and dashboarding are very essential traits that every enterprise whether big or small needs to use. The storyboards that can be painted with data are vast and can be very simple to extremely complex. We have always struggled to deliver the business insights; often business users have built their own subsystems and accomplished what they needed. But the issue is we cannot deliver all the data needed for the visualization and the dashboarding. As we progressed through the years, we have added more foundations to this visualization by adding neural networks, machine learning, and artificial intelligence algorithms. The question to answer is that the underlying data engineering has progressed and become very useable, how to harness this power into the storyboard and create the visual magic? Storyboarding is a very interesting way to describe the visualization, reporting, and analytics we do in the enterprise. We create different iterations of stories and it is consumed by different teams. How do we align the story to the requirements? What drives the insights? Who will determine the granularity of data and the aggregation requirements? What teams are needed to determine the layers of calculations and the infrastructure required?

The focus of this chapter is to discuss how to build big data applications using the foundations of visualization and storyboarding. How do we leverage and develop the storyboard with an integration of new technologies, combine them with existing databases and analytical systems, create powerful insights and on-demand analytical dashboards that will deliver immense value? We will discuss several use cases of data analytics and talk about data ingestion, especially large data sets, streaming data sets, data computations, distributed data processing, replications, stream versus batch analytics, analytic formulas, once versus repetitive executions of algorithms, and supervised and unsupervised learning and execution. We will touch specific areas on applications around call center, customers, claims, fraud, and money laundering. We will discuss how to implement robotic process automation, hidden layers of neural networks, and artificial intelligence to deliver faster visualizations, analytics, and applications.

Let us begin the journey into the world of visualization and analytics like how we saw the other chapters discuss vast data, complexity, integration, and rhythms of

Building Big Data Applications. https://doi.org/10.1016/B978-0-12-815746-6.00006-5

unstructured data processing. The new approach to visualization can be defined as the following:

- Data discovery visualization—the viewpoints of touching and seeing the raw data first, prior to deciding what is needed. In this stage, we are not looking at conceptual models or logical models of data, we are looking at identifying the data and its characteristics, its completeness, its formats, its variances, and all the different sources delivering similar data. The exercise will provide users the ability to discover data, align notes as comments, understand the business value, document operational analytics to be delivered from the data, provide data stewards guidance for the usage of the data in the data lake, data hubs, analytics, reporting, and visualization. This exercise needs to be delivered by a team of business experts on data, data analysts, and the data steward team providing the governance required for the discovery and exploration. The deliverables from this exercise will be a set of data formats, dictionaries, and metadata, and no business rules are applied at operational level. The conceptual architecture is shown in Fig. 6.1, below.
- Operational data analytics and visualization—includes the different perspectives of how data can be visualized and analyzed at an operational level. The visualization here is not applying complex business rules; it is about looking inside the data and studying the behaviors associated with the data. If you ever wanted to perform complex event processing (CEP), this is the data layer at which you can initialize the discovery, align the behaviors as you start executing the visualization, and tag the data as you see the specifics that you want to infer insights at the data lake, data hub and analytics layers. This is the stage where we will ask for a detail document to be developed for the data, its formats, its integration points, and the

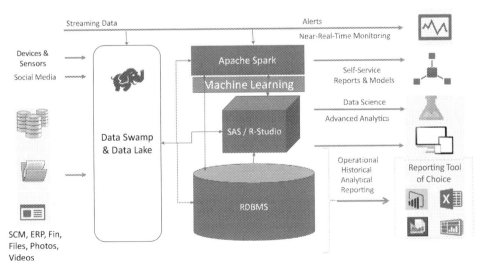

FIGURE 6.1 Conceptual data discovery and visualization architecture.

tags to retain as we process it from here to data lake. All the alignment of the data, the associated sources and the different fields and their attributions will be completed as a part of this exercise. We have struggled prior to the advent of a platform like Hadoop or Cassandra to deliver this at an enterprise scale. The argument is not whether Oracle or Teradata could not have done it, they were designed for a different purpose and the platform usage to the associated applications can be delivered only when there is a perfect alignment of the data and its usage.

• Data lake visualization—is the foundational step for enterprise usage of data for storyboarding. The foundations of the storyboard are defined in the data discovery and operational data analytics levels. The data is now loaded to an enterprise asset category and is associated with the appropriate metadata, master data, social media tags, and has integration points that will be used by different teams across the enterprise by interpreting, integrating, and visualizing the data. The visualization at this layer will resolve a lot of the foundation problems of business intelligence. The often yet another business intelligence project moniker is gone. However please understand that we need to create the technology integrations and the data engineering needs to be done. The confusion in the marketplace is the usage of the terminology data lake, which is not what a vendor can call their solution to deliver. It has to be what is built from the data insights and delivery layers. The teams involved in this data lake visualization exercise include data analytics experts, data reporting specialists, analytical modelers, data modelers, data architects, analytic visualization architects, machine learning experts, and other documentation and analysts. This team will produce documents, data models from conceptual, logical to physical, metadata, master data, and semantic data elements as identified.

Fig. 6.2 shows the data from a volumetric perspective that we need to define for the data lake creation. If this step is forgotten, please do not go beyond and ensure that it is completed. Fig. 6.3 shows the analytics and its orientation from a visualization perspective. This picture is essential to define for successful implementation of the data lake and data hubs. The analytical layers will integrate data from the operational raw data swamp all the way to data hubs, the quality of data is important as we drill-up in the hierarchy.

• Data hub visualization—is a very fine-layered data representation. This is the actual dashboard and analytics that will be used by executives and senior leaders of any enterprise. The data layer here is aggregated and summarized for every specific situation and the associated hierarchical layers are all defined, the drill-down patterns identified and aligned and the appropriate granular layers are all aligned. Today we can do this because we have visualized from the bottom most layer of data discovery all the way to analytical layers. This kind of data aggregation and integration is very unusual and has disrupted the way we do data

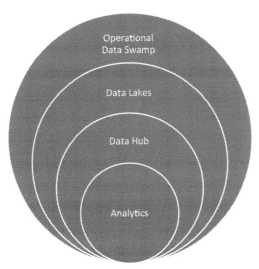

FIGURE 6.2 Data in big data applications by size.

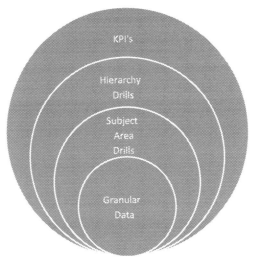

FIGURE 6.3 Analytics from visualization perspective.

management. The team required for this work is primarily data architects, analytical architects, data modelers, data stewards, and implementation teams. The team will produce document from discovery and analysis to design and code, testing and cycles of agile release plans and integrated outcomes.

- Analytics visualization—is an exercise of visualization with key performance indicators, it is a complex web of different aspects of a storyboard that need to be integrated and driven with precision. The dashboards that this layer paints present

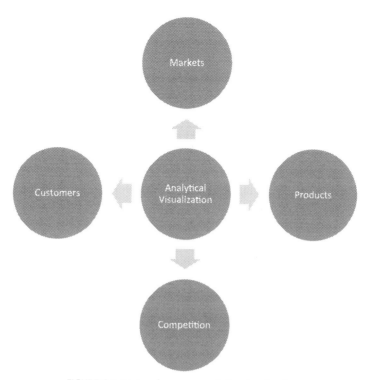

FIGURE 6.4 Porters forces on analytics visualization.

several points of interest for any executive, and usually points to critical hits or misses which reveal underlying issues to be fixed. In the world driven by internet, we need to learn this layer the fastest as the cloud can load all data and transform the same, how to bring this to life? Provide instant relief? Allow you to beat your own strategy and competition? To draw this in Porter's five forces approach, here is what we are looking at (Fig. 6.4)

The forces create a shear that keeps us riveted to understand the layers of data to provide answers as we drill down and drill across the data landscape. In this world of big data applications, it can be delivered and we will be discussing how we do this in the following segments of this chapter. Team members who participate in this include data architects, analytics modelers, data stewards, and implementation teams.

Now that segment of why visualize, what benefits does this deliver, and who does this from a team perspective, let us look at delivering the big data applications from visualizations and analytics.

The biggest impact of big data is the ability to analyze events that have happened within and outside the organization and correlate them to provide near accurate insights into what drove the outcomes. In the case of corporate data there are details that can be explored using powerful algorithms, business rules and statistical models and this data includes text, image, audio, video, and machine data. To deliver these powerful applications and their insights, we need a foundational structure for analytics and visualization of the data. Apart from the data, we need to have subject matter experts who can understand the different layers of data being integrated and what granularity levels of integration can be completed to create the holistic picture. This is the storyboard requirements to paint the application.

Big data applications programs provide a platform and the opportunity to measure everything across the enterprise. The effect of this process is the creation of transparency across the different layers of data, its associated processes and methods, exposing insights into potential opportunities, threats, risks, and issues. All users from different business teams and their executive leaders, when they get access to this application and its associated insights and data layers, gain better understanding of the decisions and are able to provide more effective guidance to the enterprise.

Big data application can be defined as the combination of visualization and data mining techniques along with large volumes of data to create a foundational platform to analyze, model, and predict the behavior of customers, markets, products or services, and the competition, thereby enabling an outcome-based strategy precisely tailored to meet the needs of the enterprise for that market and customer segment. Big data applications provides a real opportunity for enterprises to transform themselves into an innovative organization that can plan, predict, and grow markets and services, driving toward higher revenue.

Big data application storyboarding is a process of creating a narrative. The narrative will include all the specific steps of what we need to discover in the storyboard, the actual data and logs that will be used in the storyboard, the outcomes, analysis, and all identified steps of corrective actions to rerun the storyboard. The difference in this storyboard is the use of technology to rerun the process with different tweaks to check the outcomes. The experiments in research cannot be reexecuted for 100% validation, but the outcomes can be analyzed and visualized in the storyboard multiple times. Think of the experiments for cancer research or genetic chromosome sequencing studies or particle physics experiments or chemistry experiments.

The sequences of this storyboard can be altered as needed, the reason being infrastructure is very cheap and scalability is foundational, the application-specific requirements will be deliverable based on user-specific access and security. The end result of how the user reacts to the application and what they will want to explore and discover are all recordable as interactive logs and can be harnessed and replayed multiple times for visualization. This is where the new skills and delivery mechanisms are all used in the new world of data and analytics. The complexity of every visualization is documented as user interactions and each step can be broken down and managed. The formulas and calculations in the sequences are very much managed as operational data analysis and discovery to data lakes to data hubs and analytical layers are all captured and each layer will provide its own hits and misses, which means the actual analytical outcomes will be 99.9999% accurate.

Each sequence will create its own set of audit trails and the logs from these audits are very essential for the teams to have access to and analyze. The audit logs can be accessed and analyzed using Apache Flume or Apache Elastic or Splunk depending on what the enterprise wants to accomplish during the analysis. This is where we will use machine learning and neural networks the most as the logs can be varying in size, content and actions. The processing of unsupervised data is more useful to learn from without actually interfering the actual view of the execution. This flow is an essential step in creation of the actual visualization for the team.

Each audit analysis will produce the exact user interactions, system interactions and resulting steps that was the behavior of the system and the user. Think of this in research with the data from CERN particle accelerators and the discovery of the "god particle". The actual scattering of the data and its outcomes were not executable repeatedly, however if the logs are analyzed and outcomes understood, the discovery can be accelerated with the right corrections applied. This type of activity is what research teams have always wanted and have never been successful in applying their team skills together.

Let us look at some of the use cases of big data applications

Clickstream analysis—when you engage in an Internet search for a product, you can see that along with results for the product you searched, you are provided details on sales, promotions, and coupons for the product in your geographical area and the nearest 10 miles, and also promotions being offered by all internet retailers. Analyzing the data needed to create the search results tailored to meet your individual search results, we can see that the companies, who have targeted you for a promotional offer or a discount coupon. How did these retailers offer me a coupon? The answer to this is the big data analytic application in the retailer which provides details on web users' behaviors gathered with the process of clickstream analysis. The big data application is architected to provide instant analytics on any user or users' behaviors based on specific geographies or multiple geographies compared in a mash-up. To make these analytics work seamlessly and provide multiple user connects with offers, the data integration in this architecture is complex. What changes here is the actual mechanism of building this application itself.

For search to be executed, we need to have an interface that will use a search box and a submit button. This is great as all search engines provide you an interface. What happens next is the intricate process of why an engine like Google is a success while Bing or Yahoo is a distant second or third. The application development process here has to boil down to all levels of details. For search to happen, the following big data application strategy needs to be written as foundations:

- Step 0—Foundational web interface screen
  - Search text box
  - Submit button
  - Minimal logo

- Step 1—User text for search
  - User inputs a string
  - User clicks submit button
  - Search engine receives input
- Step 2—Search engine process
  - Search engine opens up metadata and semantic libraries
  - User input is evaluated for one or more strings to look for appropriate metadata
  - Based on the data evaluation of the string, the search engine will need to execute the loop of cycles to search.
  - Once the loop is determined, the string and its various combinations will be passed into the search engine
  - Each search iteration will:
    - Cycle through a metadata library
    - Match the search string with the metadata library
    - Extract all web crawls that have been indexed and stored with tags and URL addresses associated with the web crawl that the metadata matches
    - Compile a list to return to the user
    - Return the results to the loop execution logic
  - Add results to output segment
  - Return results once all loops are completed
  - The search process will do the following processes for each search:
    - Execute a neural network algorithm to perform the search and align metadata
    - Execute machine learning algorithms to align more search crawls for each execution of search, this will work with unsupervised learning techniques
    - Execute indexing and semantic extract for each search process crawl
- Step 3—Return results
  - Each search process will return a list of web URL, and associated indexed match with a rank (called page rank in Google for example)
  - Results will be spread across multiple pages to ensure all results are returned to the user.

These steps form the basic flow and stream of activities to implement and it is then developed into a set of front-end, algorithms, data search, metadata and back-end neural networks, machine learning, and web crawl. This technique is what we need to understand, the complexity in this process is in the neural networks, artificial intelligence, and machine learning processes. The expansion of these algorithms is where the search engine companies develop and deliver intellectual property and have several patents. Each web user interaction from a search result can be further recorded as clickstream and decoded on what results interested the users the most. This clickstream data when collected together can be analyzed with integrations to search terms and associated metadata. The data sets can be distributed across search applications and be leveraged across different user interfaces.

Analyzing this data further we can see that clickstream data by itself provides insights into the clicks of a user on a webpage, the page from which the user landed into the current page, the page the user visited next from the current page, the amount of time the user spent between the clicks, and how many times the user engaged in search for a category of product or service. By creating a data model to link the webpage and all these measurements, we can convert simple clicks into measurable results along with time as an associated dimension. This data can be then integrated with promotional data or campaign data to produce the offers, which are tailored to suit your needs at that point in time. Further to this basic processing, predictive analytical models can be created to predict how many users will convert from a searcher to a buyer, the average amount of time spent by these users, the number of times this conversion happened, the geographies where these conversions happened, and how many times in a given campaign such conversions can happen. While we have had similar processes created and deployed in traditional solutions, we can see by combining big data and traditional processes together, the effectiveness of the predictive analytics and its accuracy is greatly improved.

Retailers who are in the Internet and the associated ecommerce marketplace have worked on this clickstream logic for close to 2 decades, the number of techniques and algorithms have also expensed in these 2 decades. Today we are working with sensor information and mobile devices. The visualization of the information is like seeing a galaxy of data in a continuum of movements.

Another example of big data application is what you experience when you shop online today. The most popular websites offer a personalized recommendation along with products or services that you shop for. These recommendations have impacted the bottom line in a positive manner for all the eretailers who have invested in this approach. What drives a recommendation engine and how does it tailor the results to what each individual shopper searches for? If you take a step back and think through the data needed for this interaction to happen, the recommendation engine principle works on the data collected from search and purchase information that is available as clickstream and market basket data, which is harvested as lists and integrated using metadata along with geo-spatial and time information. For example, if you search for a book on "Big Data", the recommendation engine will return to you a search result and also a recommendation of popular titles that were searched when other people searched for a similar book, and additionally provides to you another recommendation on a set of books that you can buy similar to other's purchases. This data and number crunching exercise is a result set of analytics from big data.

A consumer-friendly example of big data applications is the usage of smart meters to help monitor and regulate power usage. Data from the smart meters can be read on an hourly basis and depending on the latitude and longitude coordinates, the time of the day and the weather conditions for the hour and next few hours, power companies can generate models to personalize the power consumption details for a customer. This proactive interaction benefits both the parties positively and improves the financial spend for the consumer. This type of analytics and personalized services are being implemented on a global basis by many countries.

Big data applications can be used to create an effective strategy for healthcare providers. Today whole-body scanners and X-ray machines store data electronically and this data is hosted by datacenters of large equipment manufacturers and accessed by private cloud services. Harnessing the infrastructure capabilities, different providers can study the patient data from multiple geographies and their reviews can be stored electronically along with recommended treatment options. This data over a period of time can be used as an electronic library of treatments, disease state management, and demographic data. We can create heatmaps on the most effective treatments and predict with confidence what therapies are effective for regions of the world, and even extend this to include gene therapies and genetic traits based on regions of the world and ethnicity. Doctors can derive propensity analytics to predict outcomes and insurance companies can even harvest this data to manage the cost of insurance.

As we see from these examples, big data applications create an explosion of possibilities when the data is analyzed, modeled, and integrated with corporate data in an integrated environment with traditional analytics and metrics.

The next segment of big data application is to understand the data that we will leverage for the applications. The data platform is a myriad of types of data and it includes the following: (Fig. 6.5)

This data platform is discovered in the data discovery exercise and all the operational data can be consumed as needed by every user. In developing big data applications, the process of data discovery can be defined as follows:

- Data acquisition—is the process of collecting the data for discovery. This includes gathering the files and preparing the data for import using a tool. This can be accomplished using popular data visualization tools like Datameer, Tableau, and R.

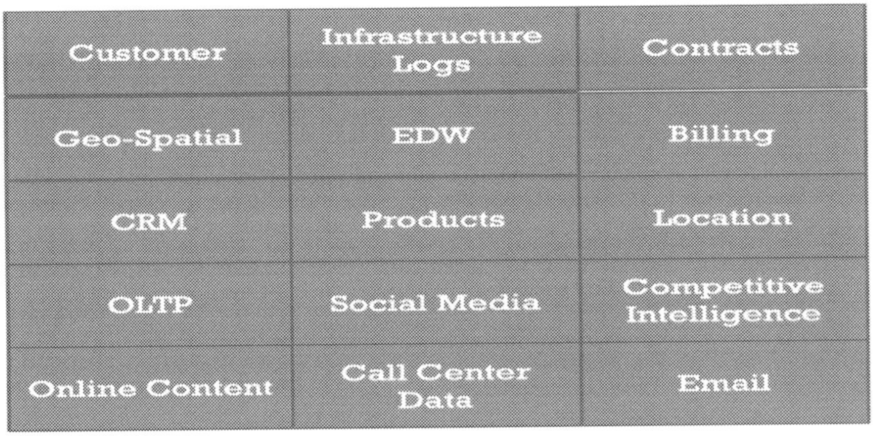

Customer	Infrastructure Logs	Contracts
Geo-Spatial	EDW	Billing
CRM	Products	Location
OLTP	Social Media	Competitive Intelligence
Online Content	Call Center Data	Email

FIGURE 6.5 Data types used in big data applications.

- Data tagging—is the process of creating an identifying link on the data for metadata integration.
- Data classification—is the process of creating subsets of value pairs for data processing and integration. An example of this is extracting website URL in clickstream data along with page-view information.
- Data modeling—is the process of creating a model for data visualization or analytics. The output from this step can be combined into an extraction exercise.

Once the data is prepared for analysis in the discovery stage, the users can extract the result sets from any stage and use it for integration. These steps require a skill combination of data analytics and statistical modeling, which is the role of a data scientist. The question that confronts the users today is how to do the data discovery, do you develop MapReduce code extensively or do you use software like Tableau or Apache Presto. The answer to this question is simple, rather than develop extensive lines of MapReduce code, which may not be reusable, you can adopt to using data discovery and analysis tools that actually can produce the MapReduce code based on the operations that you execute.

Depending on whichever method you choose to architect the solution your data discovery framework is the key to developing big data analytics within your organization. Once the data is ready for visualization, you can integrate the data with mash-ups and other powerful visualization tools and provide the dashboards to the users.

Visualization

Big data visualization is not like traditional business intelligence where the data is interactive and can be processed as drill downs and roll-ups in a hierarchy or can be drilled into in a real-time fashion. This data is static in nature and will be minimally interactive in a visualization situation. The underlying reason for this static nature is due to the design of the big data platform like Hadoop or NoSQL, where the data is stored in files and not in table structured, and processing changes will require massive file operations, which are best, performed in a microbatch environment as opposed to a real-time environment. This limitation is being addressed in the next generation of Hadoop and other big data platforms.

Today the data that is available for visualization is largely integrated using mash-up tools and software that support such functionality including Tableau and Spotfire. The mash-up platform provides the capability for the user to integrate data from multiple streams into one picture, by linking common data between the different datasets.

For example, if you are looking at integrating customer sentiment analytics with campaign data, field sales data, and competitive research data, the mash-up that will be created to view all of this information will be integrating the customer sentiment with campaign data using product and geography information, the competitive research data and the campaign data by using geography information, the sales data and the campaign data by using the product and geography information. This dataset can be queried interactively and can be used for what-if type of causal analysis by different users across the organization.

Another form of visualization of big data is delivered through the use of statistical software such as R, SAS, and KXEN, where the predefined models for different statistical functions can use the data extracted from the discovery environment and integrate the same with corporate and other datasets to drive the statistical visualizations. Very popular software that uses R for accomplishing this type of functionality is RStudio.

All the goods that we are discussing in the visualization can be successfully completed in the enterprise today, with the effective implementation of several algorithms. These algorithms will be implemented as portions of formulation and transformation of data across the artificial intelligence, machine learning, and neural networks. These different implementations will be deployed for both unsupervised learning and supervised learning, and we will benefit in visualization from both the techniques. The algorithms include the following and several proprietary implementations of similar algorithms within the enterprise.

- Recommender
- Collocations
- Dimensional reduction
- Expectation maximization
- Bayesian
- Locally weighted linear regression
- Logistic regression
- K-means clustering
- Fuzzy K-means
- Canopy clustering
- Mean shift clustering
- Hierarchical clustering
- Dirichlet process clustering
- Random forests
- Support vector machines
- Pattern mining
- Collaborative filtering
- Spectral clustering
- Stochastic singular value decomposition

The teams in the enterprise for this visualization and associated algorithms are the teams of the data scientist.

The evolving role of the data scientist

There is a new role in the world of data management that has evolved with big data called a "data scientist". There are several definitions for this role that are evolving including data analysts with advanced mathematical degrees, statisticians with multiple specialist degrees, and much more. In a simple language speak, the "data scientist" is a role where the person has intimate knowledge of the data being discovered and can create effectively explore the data and infer relationships that create the foundation for analytics and visualization.

The key role that enables the difference between success and failure of a big data program is the data scientist. The term was originally coined by two of the original data scientists DJ Patil and Jeff Hammerbacher when they were working at LinkedIn and Facebook.

What defines a data scientist? Is this a special skill or education? How different are these roles from an analyst or engineer?

There is no standard definition for the role of a data scientist, but here is a close description—A data scientist is an expert business analyst or an engineer who uses data discovery tools to find new insights in data by using techniques that are statistical or scientific in nature. They work on a variety of hypothesis and design multiple models that they experiment with to arrive at new insights. To accomplish this, they use a large volume of data, which is collectively called as big data.

Data scientists work very closely with data and often question everything that is input or output from the data. In fact, in every enterprise there are a handful of senior business analysts or data analysts that are playing the role of the data scientist without being formally called as one.

Data scientists use the data discovery tools discussed in this chapter to create the visualization and analytics associated with big data. This role is still in evolution phases and in the future we will see many teams of data scientists in enterprises as opposed to a handful that we see today. If data is the new oil, then the data scientist is the new explorer.

In summary we can use big data to enhance analytics and deliver data for visualization and deeper analysis as needed by the enterprise. There are evolving techniques, methods, and technologies to accomplish this on a regular basis within enterprises. The underlying goal that is delivered with big data analytics is the capability to drive innovation and transformation across the enterprise in a transparent and informed manner, where you can tie the outcomes to the predictions and vice versa. The possibilities are endless and can be delivered from the same dataset as it is transformed in discovery processes. This kind of a flexible approach is what you should adapt to when designing the solutions for big data analytics and visualization.

7

Banking industry applications and usage

Digital transformation has changed banking largely across the world. We are seeing the disappearance of the ATM, the emergence of open ledger distributed processing, blockchain, crypto currencies, and consumers who do not even feel the need for a physical bank. How will we deliver to the consumer who is digital? This is where big data applications come to focus on banking industry. The connected consumer today has mobile devices to be transacting always, we will discuss how to develop and connect with that consumer with applications that provide constant value. The goal is to deliver monetization and increase efficiency of business, while reducing risks and avoiding customer churn. What can be done and how to use data to accomplish the goal will be the focus of the applications. The effective storyboard creation, the analytics and charts, the visualization, user communication, and dashboards.

The era of electronic data processing started back in the 1960s. The foundations were laid for the generation, storage and processing of large volumes of data. However, the mass data era only began with the expansion of the internet since the 1990s, the digital world has become part of the daily life of consumers, leading to rapid data growth. In addition, rapid changes and evolution of mobile computing and phones have created a massive increase in image, sound, and position data. Today, the Internet of Things (IoT) is increasing the level of connectivity and has given rise to a growing number of measuring points via sensors over the last years. This concept was discovered in 1800s by studying the impact of crowdsourcing, where the connected community of users would make a collective decision which is very close to the actual requirement or the actual values. All of these aspects are applied today with the implementation of AI and Machine learning. The millennial age customer is a different persona, and their expectations for any service is speed of "google like" delivery, with all associated information and details being accessible for further analysis, at any time and any moment of requirement. This customer is transforming the foundations of retail, healthcare, banking, insurance, financial services, and education at the least expressed.

The complexity of this new generation is amplified with the amount of activity, the volume of data generated, the continuous monitoring and alerts they expect, the nuances of their expressions, and the crowd that is created, participated, and followed by this customer or prospect. All of this has tremendous business value that can be harnessed through the creation of applications with smart features built in and available on demand. This expectation is not new, we have evolved our process and aligned with available technology frontiers, however we have reached the tipping point of

Building Big Data Applications. https://doi.org/10.1016/B978-0-12-815746-6.00007-7

infrastructure with the innovation of NoSQL databases and Hadoop, both of them enriching the available layers with the bottom-most and top-most layers of infrastructure. In this new world, the implementation of AI-driven process engineering and ability to execute as independent automated processes is where we bring in complex event processing and operations algorithms which are very useful when implemented. Fig. 7.1 shows the entire thought encapsulated as a user experience.

Crowd engineering is a process engineering model which connects multiple unrelated processes, which are actually related and have financial impacts. Look at online shopping for example, a consumer searches for products, finds what they are looking for and either buy it or walk away from the entire process. The expectation is that one either made a purchase or one did not, what if there were cookies that watched the entire process, stole the credit card information for the consumer, and then transacted without their knowledge to make purchases. The consumer is alerted by the bank that manages the credit card and eventually is protected by FDIC rules for fraud. However, the banks have to write-off the transaction using their fraud insurance. This issue needs to be managed better and nobody should have to bear the consequences of fraud, but how can we get to that end state? This is where the application of AI and machine learning will help. The front-end browser and the associated financial pages can be encrypted with algorithms that make it useless for a cookie or even an advanced persistent threat to steal information. Have we done it? Yes or no, this is a later discussion.

AI and machine learning are having a major impact on banking, driven by vast processing power and the continual development of new and more accessible tools and, of course, the sheer volume of potentially useful, accurate data that all banks offer today. Both retail and investment banks use machine learning in numerous contexts, from the relatively mundane business of approving loans, to fund management, risk assessment, and more. It is true that banks have used computer algorithms to trade stocks and shares for decades. But that started when machines learning was more in a lab mode of innovation. Today applications of machine learning in banking involve understanding social media, news trends, and other data sources—beyond stock prices and trades.

Today, an algorithm can play a part in calibrating a financial portfolio to the precise goals and risk tolerance of the user, ensuring ideally, that a defined amount is earned by a certain date from money invested. It can even autonomously adjust the management of portfolios as market conditions change. An intelligent system scours millions of data points, including granular trading information on companies around the world. It then comes up with moneymaking strategies that it executes, as neural networks with minimal supervision.

FIGURE 7.1 Crowd engineering.

AI and big data for banking applications autonomously spot patterns humans cannot see. It can also self-correct to keep on improving and adapting to changes in the market. Which brings us to the term that comes up again and again in machine learning and AI personalization.

Personalized drug treatments, ad campaigns, and now an AI system that can help a bank create and recommend better banking products to customers on a personalized basis. Convincing a customer to plan or provide them with decisions already made for them are both tricky and will need to learn over time. AI and machine learning can help marketers to target high potential customers at a time when they are most likely to respond favorably. However, one application that will no doubt excite a lot of interest among banks and customers is fraud detection. Of course, an unwary individual could lose a few thousand dollars by being too relaxed with their bank details, but an institution, with an increasing amount of valuable company data being stored online, could be a victim on a much grander scale. And this is where machines really do learn: by recognizing established threats or potential threats and adjusting to new ones.

Technology implementation within applications today uses AI techniques to analyze internal, publicly available, and transactional data within a customer's wider network to spot rogue behavior has been piloted. Cybersecurity threats have gotten more complex, clever, and quick, and machine learning that can adapt will be invaluable, where we see applications including Apache Metron and Cisco Umbrella. Facial recognition, voice recognition, or other biometric data can be added to market movements and transactions to develop a data mountain accessible and understandable by machines that can reveal patterns that may be threatening. These data sources, of course, have applications not only to security but to customer relations and marketing. These are all trends of what is happening within banking today.

The multi-billion-dollar parts of finance: loans, insurance and underwriting are highly competitive businesses, and the information to be processed and managed to make these segments effective is complex and challenging. This is another interesting application where machine learning and neural networks have transformed the processing, increased the availability of data, and associated process reduced time to process and have improved the data lineage and overall process.

Age, health, and life expectancy are a changing constant and they have all imposed requirements that need to be managed and adjusted along the way, but how about looking at the habits of a certain age group in a certain area over a long period to assess otherwise unnoticed risks or benefits? Do they drink too much in one county? Do they live in a city with access to medical marijuana? Do they spend more time on computers or mobile devices online? Have healthier lifestyles? Manage their money better than a similar group in another town? How precisely can these elements be assessed, calibrated, and used? And can they be scaled up to millions of examples of consumer data and then applied accurately to insurance risks over an entire population? These are all questions that have been a tiger's tail chase for banks. Today with the emergence of algorithms, neural networks, and natural language processing (NLP) applications, in the backend infrastructure we constantly attempt to understand natural human communication, either written or spoken, and communicate in return with humans, using similar, natural language. AI and machine learning help organizations create consistency and a personalized experience across channels over time and give them a competitive edge.

None of this is net new except the changes in innovating the underlying infrastructure, improved efficiency of streaming processing layers, deploying extremely faster networks, and innovating cybersecurity and distributed data processing with neural networks.

The coming of age with uber banking

In 2015, an article appeared in Washington Post which read like this, "The neighborhood bank branch is on the way out and is being slowly phased out as the primary mode of customer interaction for Banks. Banks across the globe have increased their technology investments in strategic areas such as Analytics, Data & Mobile. The Bank of the future increasingly resembles a technology company". Technology is transforming Banking and it is leading to dramatic changes in the landscape of customer interactions. Today we live in the age of the hyper-connected consumer. As millennials enter and engage in banking, they are expecting to be able to Bank from anywhere, be it a mobile device or use internet banking from their personal computer, at all times and have access to all services.

As former Barclays CEO Antony Jenkins described it in a speech the global banking industry, is under severe pressure from customer demands for increased automation and contextual services. He said "I have no doubt that the financial industry will face a series of Uber moments," he said in *the late-November speech in London, referring to the way that Uber and other ride-hailing companies have rapidly unsettled the taxi industry*. **The outcome of these mounting pressures over the last 8 years from 2010, has led to banking trends migrating to becoming more contextual, social, and digital to respond to changing client needs.**

The financial services industry segments including banks, mortgage, insurance, and other associated industries have been facing an unprecedented amount of change driven by factors like changing client preferences and the emergence of new technology—the Internet, mobility, social media, etc. These changes are immensely impactful, especially with the arrival of "FinTech"—technology-driven applications that are upending long-standing business models across all sectors from retail banking to wealth management and capital markets. The new market of a major new segment, millennials, use mobile devices, demand more contextual services, and expect a seamless unified banking experience, something similar and aligned to what they experience on web properties like Facebook, Amazon, Uber, Google or Yahoo, etc.

A true digital bank needs to identify all the key areas where itis capabilities will need transformation to occur. Some of these include the following:

- Offer a seamless customer experience much like the one provided by the likes of Facebook and Amazon, a highly interactive and intelligent applications stack that can detect every customer's journey across multiple channels
- Offer data driven interactive services and products that can detect customer preferences on the fly, match them with existing history and provide value added

services. Services not only provide a better experience but also foster a longer-term customer relationship and earn trust and loyalty eventually.
- Ideate and develop new business prototype, test, refine and rapidly develop, design, and deploy business capabilities
- Transform the vision and mission statement to become digital as a **constant capability** and not as an "off-the-shelf" product or a one-off way of doing things.

These are some foundational thoughts and today all banks of any size and shape have transformed their service portfolio and are emerging more digital than ever before. Add the emergence of cryptocurrency, blockchain, and more open-ledger transformations, we are looking at a completely different future for banking. This is the tipping point of FinTech and its evolution to digitize the entire spectrum and create a future bank.

FinTech's (or new age financial industry startups) offer a different model of customer experiences designed and developed on customer on product innovation and agile business models. The innovation is delivered by reducing some profit shares but instead they benefit by getting more usage and activity, all of this focused around contextual products tailored to individual client profiles. The increased ability of business subject matter experts in analytics and, the savvy use of segmentation data and predictive analytics enables the delivery of bundles of tailored products across multiple delivery channels (web, mobile, point of sale, Internet, etc.).

Compliance is mandatory requirement for banks in areas around KYC (know your customer) and AML (antimoney laundering) where there is a need to profile customer both individual and corporate to decipher if any of their transaction patterns indicate money laundering, etc. Due to these compliance requirements, banking produces the most data of any industry that pertain to customer transactions, payments, wire transfers, and demographic information. However, it is not enough for financial service IT departments to just possess the data. They must be able to drive change through legacy thinking and infrastructures as the industry changes—both from a data product as well as from a risk and compliance standpoint. FinTech's are not mandated to be regulatory compliant and the technologies support multiple modes of payments at scale, however the banking regulatory bodies mandate that when a FinTech is brought into the bank, it must pass all compliance requirements.

The business areas shown in Fig. 7.2, are a mix of both legacy capabilities (risk, fraud, and compliance) to the new value-added areas (mobile banking, payments, omnichannel wealth management etc).

Business challenges facing banks today

Banks face business challenges across **three** distinct areas. First, they need to play **defense** with a myriad of regulatory and compliance legislation across areas such as risk data aggregation and measurement and financial compliance, and fraud detection. Second, there is a distinct need to **improve customer satisfaction** and increase loyalty by implementing predictive analytics capabilities and generating efficient insights across the customer journey driving a truly immersive digital experience. Lastly, banks need to

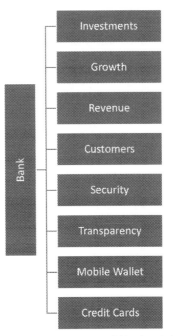

FIGURE 7.2 Predictive analytics and big data in banking.

leverage the immense mountains of data assets to **develop new business models**. The strategy to do this is by monetizing multiple data sources both data-in-motion and data-at-rest for actionable intelligence.

Data is the single most important driver of customer transformation, impacting financial product selection, promotion targeting, next best action and ultimately, the entire consumer experience. Today, the volume of this data is growing exponentially as consumers increasingly share opinions and interact with an array of smart phones, connected devices, sensors, and beacons emitting signals during their customer journey. Business and technology leaders are struggling to keep pace with a massive glut of data from digitization, the Internet of Things, machine learning, and cybersecurity.

To minimize the negative downside of data analytics and deliver better results, we today have big data-driven predictive analytics. The Hadoop platform and ecosystem of technologies have matured considerably and have evolved to supporting business critical banking applications. Align this with the ability to integrate cloud computing, mobile and device-driven engagement models and social media integration, we will create new opportunities and leverage the analytics to the best extents possible.

Positively impacting the banking experience requires data, which is available and we need to orchestrate models of usage and analytics to leverage the data

- **Retail and consumer banking**—Banks need to move to an online model, providing consumers with highly interactive, engaging, and contextual experiences that span multiple channels across eBanking. Further goals are increased profitability per customer for both microcustomerand macrocustomer populations with the ultimate goal of increasing customer lifetime value (CLV). This needs to be planned with reduction of physical locations like ATM or Service branches, which drain into the existing profit margins. The eBanking range of services will generate more data and activity tracking which can be completed successfully.

- **Capital markets**—Capital market firms need to create new business models and offer more efficient client relationship services based on their data assets. Those that leverage and monetize their data assets will enjoy superior returns and raise the bar for the rest of the industry. The most critical step is the maturity journey with analytics is to understand all their clients from a 360-degree perspective so they can be marketed to as a single entity across all channels of engagement, which is a key to optimizing profits with cross-selling in an increasingly competitive landscape.
- **Wealth managers**—The wealth management segment (e.g., private banking, tax planning, estate planning for high net worth individuals) is a potential high growth business for any financial institution. It is the highest touch segment of banking, fostered on long-term and extremely lucrative advisory relationships. It is also the segment most ripe for disruption due to a clear shift in client preferences and expectations for their financial future. Actionable intelligence gathered from real-time transactions and historical data becomes a critical component for product tailoring, personalization and satisfaction.
- **Corporate banking**—The ability to market complex financial products across a global corporate banking client base is critical to generating profits in this segment. Analytical models will be useful to engage in risk-based portfolio optimization to predict which clients are at risk for adverse events like defaults. In addition to being able to track revenue per client and better understand the entities they bank with, we need to provide the compliance needs to track AML, which is an international mandate.

The customer journey and the continuum

Across retail banking, wealth management and capital markets, a unified view of the customer journey is at the heart of the bank's ability to promote the right financial product, recommend a properly aligned portfolio products, keep up with evolving preferences as the customer relationship matures and accurately predict future revenue from a customer. Today this aspect of integration of the customer and providing a single view across all channels of engagement has been adopted and become a key transformation in banking. We can easily visualize the activities across connected accounts and all investment, credit cards, and manage wire transfers and payments from one application. Leveraging the ingestion and predictive capabilities of a big data platform, banks can provide a user experience that rivals Facebook, Twitter, or Google and provide a full picture of customer across all touch points. Is all this data connected seamlessly at the backend? The evolution is happening and will be a new standard by 2025.

Create modern data applications

Banks, wealth managers, stock exchanges, and investment banks are companies run on data generated by customer during activity including deposits, payments, balances, investments, interactions, and third-party data quantifying risk of theft or fraud. Modern data applications for banking data scientists preferably need to be built internally as they can be customized on-demand or can be implemented by a purchase from "off-the-shelf" third parties. The ability to add artificial intelligence and machine learning algorithms have delivered the new applications are powerful and fast enough to detect previously invisible patterns in massive volumes of real-time data. They also enable banks to proactively identify risks with models based on petabytes of historical data. These data science apps comb through the "haystacks" of data to identify subtle "needles" of fraud or risk not easy to find with manual inspection.

The use cases of analytics and big data applications in banking today

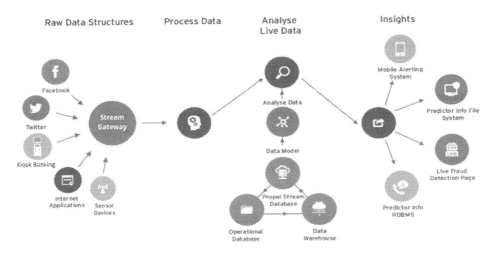

Understanding your customer interactions.

The customer buying journey requires an extensive optimization of the customer acquisition and loyalty campaigns; you need greater visibility into the customer buying journey as it transcends different segments of your business. We need to collect and integrate analytics across all these activities to improve the journey.

- Increase customer acquisition
- Increase revenue per customer
- Decrease customer acquisition cost

- Reduce customer churn
- Product enhancement

Deeper, data-driven customer insights are critical to tackling challenges like improving customer conversion rates, personalizing campaigns to increase revenue, predicting and avoiding customer churn, and lowering customer acquisition costs. And consumers today interact with companies through lots of interaction points – mobile, social media, stores, e-commerce sites, and more dramatically increasing the complexity and variety of data types you have to aggregate and analyze. Think web logs, transaction, and mobile data, advertising social media and marketing automation data, product usage and contact center interactions, CRM and mainframe data, and publicly available demographic data. When all of this data is aggregated and analyzed together, it can yield insights you never had before—for example, who are your high-value customers, what motivates them to buy more, how they behave, and how and when to best reach them. Armed with these insights, you can improve customer acquisition and drive customer loyalty.

The application for this process will be built using the big data analytics platform and leverage all the data—structured and unstructured, because you can combine, integrate, and analyze all of your data at once to generate the insights needed to drive customer acquisition and loyalty. The process can be customized to deliver different types of insights for different teams of users or even a single management executive. For example, you can use insights about the customer acquisition journey to design campaigns that improve conversion rates. Or you can identify points of failure along the customer acquisition path or the behavior of customers at risk of churn to proactively intervene and prevent losses. And you can better understand high-value customer behavior beyond profile segmentation, including what other companies they shop from, so you can make your advertisements even more targeted.

Fraud and compliance tracking

Today in the world we live in, if you are responsible for security, fraud prevention, or compliance, then data is your best friend. You can use it to identify and address issues before they become problems. The fact is, security landscapes and compliance requirements are constantly evolving, as are the methods that the bad guys are using to defraud your business and customers. What can be delivered as applications include the following/;

- Cyber attack prevention
- Regulatory compliance
- Criminal behavior
- Credit card fraud detection

Data-driven insights can help you uncover what's hidden and suspicious, and provide you insights in time to mitigate risks. For example, analyzing data can help you reduce the operational costs of fraud investigation, anticipate and prevent fraud, streamline regulatory reporting and compliance (for instance, HIPPA compliance), identify and stop rogue traders, and protect your brand. But this requires aggregating and analyzing data from a myriad of sources and types and analyzing it all at once, not a small task. Think financial transaction data, geo-location data from mobile devices, merchant data, and authorization and submission data. Throw in data from lots of social media channels and your mainframe data, and you have a significant challenge on your hands. However, with the right tools, this melting pot of data can yield insights and answers you never had before, insights you can use to dramatically improve security, fraud prevention, and compliance.

The applications here will leverage big data analytics and combine, integrate, and analyze all of the data at once, regardless of source, type, size, or format, and generate the insights and metrics needed to address fraud and compliance-related challenges. For example, you can perform time series analysis, data profiling and accuracy calculations, data standardization, root cause analysis, breach detection, and fraud scoring. You can also run identity verifications, risk profiles, data visualizations, and perform master data management.

Client chatbots for call center

Chatbots are virtual assistants that are designed to help customers search for answers or solve problems. These automated programs use NLP to interact with clients in natural language (by text or voice), and use machine learning algorithms to improve over time. Chatbots are being introduced by a range of financial services firms, often in their mobile apps or social media. The evolution is in a stage 2 for the chatbots, the potential for growth has seen an increasing usage, especially among the younger generations, and become more sophisticated. The stage 1 generation of chatbots in use by financial services firms is simple, generally providing balance information or alerts to customers, or answering simple questions. The next generation chatbots in the banking industry will revolutionize the customer relationship management at personal level. Bank of America plans to provide customers with a virtual assistant named "Erica" who will use artificial intelligence to make suggestions over mobile phones for improving their financial affairs. Allo, released by Google is another generic realization of chatbots.

Antimoney laundering detection

Antimoney laundering (AML) refers to a set of procedures, laws, or regulations designed to stop the practice of generating income through illegal actions. In most cases, money launderers hide their actions through a series of steps that make it look like money that came from illegal or unethical sources are earned legitimately. In a striking move to detect and prevent AML from occurring, most of the major banks across the globe are shifting from rule-based software systems to artificial intelligence designed and imple-mented systems which are more robust and intelligent to the antimoney laundering patterns. As we evolve the systems and mature their actions, these systems are only set to become more and more accurate and fast with the continuous innovations and im-provements in the field of artificial intelligence.

Algorithmic trading

Plenty of banks globally are using high end systems to deploy artificial intelligence models which learn by taking input from several sources of variation in financial markets and sentiments about the entity to make investment decisions on the fly. Reports claim that more than 70% of the trading today is actually carried out by automated artificial intelligence systems. Most of these investors follow different strategies for making high frequency trades (HFTs) as soon as they identify a trading opportunity based on the inputs. A few active banks and financial services companies in AI space are: Two Sigma, PDT Partners, DE Shaw, Winton Capital Management, Ketchum Trading, LLC, Citadel, Voleon, Vatic Labs, Cubist, Point72,and Man AHL.

Recommendation engines

Recommendation engines are a key contribution of artificial intelligence in banking sector. It is based on using the data from the past about users and/or various offerings from a bank like credit card plans, investment strategies, funds, etc. to make the most appropriate recommendation to the user based on their preferences and the users' history.

Recommendation engines have been very successful and a key component in revenue growth accomplished by major banks in recent times. Implementing the bg data platforms and faster computations, machines coupled with accurate artificial intelligence algorithms are set to play a major role in how recommendations are made in banking sector.

Cognitive search represents a new generation of enterprise search that uses sophis-ticated algorithms to increase the relevance of returned results. It essentially moves the nature of searches from basing relevance on keyword hits to understanding user intent, observing behaviors, and applying pattern detection to correctly assert the most relevant pieces of information. Structuring data and finding relations within it can bring tremendous additional business value. Even more, value can be created by employing smart analytic tools in combination with machine learning.

Training these algorithms with the valuable expertise of analysts can be a game changer that allows a bank to differentiate itself and lead to even more educated in-vestment decisions. The tools to harvest the full potential of data are here today.

Search results being intuitive and cognitive will provide you with filtering and dashboarding solutions allowing the generation of summarized information from various sources, in turn enabling users to dive deep into the data for efficient and insightful exploration. Users gain 360-degree views on what is happening, why it is happening, what to do next, and who should be involved in the process.

These modern data applications make big data and data science ubiquitous. Rather than back-shelf tools for the occasional suspicious transaction or period of market volatility, these applications can help financial firms incorporate data into every decision they make. They can automate data mining and predictive modeling for daily use, weaving advanced statistical analysis, machine learning, and artificial intelligence into the bank's day-to-day operations.

A strategic approach to industrializing analytics in a banking organization can add massive value and competitive differentiation in five distinct categories:

1. Exponentially improve existing business processes. eg., Risk data aggregation and measurement, financial compliance, fraud detection
2. Help create new business models and go to market strategies—by monetizing multiple data sources—both internal and external
3. Vastly improve customer satisfaction by generating better insights across the customer journey
4. Increase security while expanding access to relevant data throughout the enterprise to knowledge workers
5. Help drive end-to-end digitization

Predicting customer churn

Let us apply the learnings discussed to an application for predicting customer churn, which is a big drain on resources and finances in banks. The cost of acquiring a customer is higher than the cost of retaining a customer and all banks try to protect themselves against an increase in the churn rate, using customer behavior analysis providing an early indicator for banks. In order for the customer behavior to work effectively, we need to get a 360-degree global view of each customer and their interactions on different exchange channels such as banking visits, customer service calls, web transactions, or mobile banking. The transactional behavior will provide warning signs, such as reducing transactions or cancellation of automatic payments. However, increasing the volume, variety, and velocity of the data needing to be exploited has made it nearly impossible to store, analyze, and retrieve useful information through traditional data management technologies.

Big data infrastructure has been designed and built to address these challenges by solving data management problems by storing, analyzing, and retrieving the massive volume and variety of structured and unstructured data while scaling with extreme flexibility and elasticity as data increases. The platform and its application will provide banks benefit from real-time interactions with their customers.

How does this work from an application perspective? Banking applications have all focused on customer and their journey since banking started as an enterprise. We have always ensured that customer is the king and all services provided to customers have been validated as being valuable. The bank evolved into being a part of the family for the customer and their financial expectations from loans to investments and retirements are all managed by the bank. The application requirements from the customer perspective include the following:

- Access to all accounts, balances, and transactions with up to date status
- All transactions and associated details
- All credit card accounts
- All loan accounts
- All investment accounts
- All household information

The expectation is one application interface with access, security, device support, interactive call center support, and more robotic assistance with real person support as needed. Where do we lose sight of this customer and their expectations? How do we proactively engage with this opportunity to turn it to the advantage of the bank while benefiting the customer? This is where the application development for predicting churn will take form and shape.

The basic requirements for this application include the following:

- Dashboard—a central predictive analytics dashboard that can be shared across call centers, back office, and agents in the bank
- Customer engagement window—this is the logging and tracking of every activity the customer engages with the bank either through the application on a computer, mobile device, or in-person with an agent. The logging happens as the applications are used for conducting transactions and the average length of time each activity occurred. The measure of time is critical here and cannot be missed.
- Sentiment collection—is a key activity to complete for each interaction, it is not a long-haul list of questions or survey, it has to be a smiley face or a thumbs up emoji that is needed to be collected. If the customer is dissatisfied they might not reply or we need to watch for a social media outburst of poor service in the next few hours.
- AI and machine learning algorithms—the data collected in the logging and the sentiment response will be processed using AI and ML algorithms in a non-connected backend process. The algorithm will identify the customer and their transactions in a private container, and add identities of bank agents if they were involved in the process. The algorithms will look at behavior, time, transaction types, number of similar transactions executed, time in days between each trans-action, online versus offline transaction, sentiment of customer, and associated outcomes. This data will be then reviewed based on potential prediction outcomes and impact analytics will be drawn for this customer. A grouping of similar cus-tomers, their geographies, age demographics, and earning demographics will be calculated and predicted for churn.

- Outcomes—will be updated with graphics based on outcomes from algorithms, with alerts being sent to managers, and administration teams on critical churn outcomes.

Let us define customer churn and see how we will deliver the application. Definition of churn is the movement of customer from one bank to another. The reasons for example be:

- Availability of latest technology in the competitor
- Customer-friendly bank staff and minimal interference for any service from staff
- Low interest rates in loans and credit cards
- Location of the physical branch close to home or work
- Services offered in mobile with blockchain, presenting so many new ways to interact with the bank
- Churn rate usually lies in the range from 10% up to 30%

Why does churn happen? Customer dissatisfaction is a primary reason and how do banks invite this behavior from their customers, some reasons include the following:

1. **Fees**—Raising fees on financial service products is the number one reason why customers consider leaving in the first place. However, poor service ends up being the primary reason for actually leaving. Offering lower fees to match online competitors is a difficult initiative for banks looking for short-term growth, but there are other levers you can pull. Customers appreciate it when they feel taken care of, and a customer experience impact study found that 86% of consumers are willing to pay more for a better customer experience.
2. **Rates**—Much like fees, the return rates offered by products like money market funds and savings accounts are highly competitive across finTech startups, new online entrants, and the incumbents. By implementing a long-term strategy to offer lower fees on entry-level products and better rates on more profitable products (e.g., mortgage and auto loans), companies can increase the customer lifetime value and increase loyalty.
3. **Branch locations**—Some institutions have adopted an online-only model recently to save costs, but the majority of customers still like having a physical bank to get in-person guidance from a valued advisor on investment strategies. A study and report by Qualtrics found that 73% of online-only customers said they would switch to a brick-and-mortar bank, while only 40% of brick-and-mortar customers affirmed they would switch to an online-only alternative.
4. **24/7 customer service and wait times**—No one wants to be on hold for any amount of time, especially when there is financial risk involved. If your wait times are at or above the industry average, consider implementing an online chatbot to address easy questions and a guaranteed call-back system from a live support rep for more complex issues. Around the clock live support is expensive, but most

customers can be satisfied as long as there is some option to raise an issue, whether in-person, online, or over the phone.

5. **Breadth of product offerings**—Customers want options and to feel they received the best rate when shopping for big-ticket items that require financing. Qualtrics research found that many consumers maintain checking and savings accounts at a primary bank, but shop around for better rates on mortgages, car loans, investments, and credit cards. Deploying a strategy to offer competitive rates on long-term, higher profit investment vehicles, your personal relationship can help keep your customer's finances in-house.

6. **Quality of digital tools**—In a recent study on customers of banks by Qualtrics, all customers surveyed spend 69% of their time online or using the mobile app, with millennials even higher at 79%. As more of our daily needs can be accomplished with a smartphone, you must ensure your online and mobile experiences are smooth, easy, secure, and enjoyable for managing one's finances. Adding educational resources about financial investments and other nonconventional services is also a valuable option to digital customers.

The services across these areas are in a constant state of improvement; however, we need to develop the entire spectrum of solutions to provide relevant applications. These applications need variables that can be tracked and implemented, these include the following:

- Four types of variables that are mostly used:
  * Customer demographics variables
    - Age
    - Job type
    - Gender
    - Family size
    - Geographical data
  * Perceptions variables—Perception variables try to measure how the customer appreciates the services or the products, they Include dimensions such as:
    - Quality of services
    - Satisfaction with the company
    - Locational convenience
    - Pricing
  - Behavioral variables—Behavioral variables look at the previous behavior of the customer
    - How often they use a service or product
    - Which services or products is used the most and least by customer, geography, gender, employed status, student, and more.
    - Most popular behavioral variables are number of purchases and money spent

- Macroenvironment variables—Macroenvironment variables focus on identifying changes in the world that could affect the customer.
- The churn variable that we will use for our application and dashboard include
  - Demographic variables and behavioral variables
  - All other variables that are available in the bank database

Now the definition of churn and associated modes of activities are clear, we need to define how we will engage in tracking the engagement windows and its outcomes. Engagement windows for customer activity include

- Online banking
- Internet shopping
- Wire transfers
- Call center activity
- Campaign response
- Computer banking
- In−branch banking

Each of these activities generate several logs for customer, activity, channel, time of engagement, model of activity, sentiment expressed across social media, and influencer and follower models of activity. Additional data can be purchased today for similar activity across competitors and in regulatory compliance, the actual username and sensitive data if any are not shared. The next step is to identify the algorithms that can process this data independently and can harness and tether data once processed to align and provide insights as needed by the users.

Algorithms to be used in artificial intelligence and machine learning models will include the following

- Naïve Bayes Classifier Algorithm
- K Means Clustering Algorithm
- Support Vector Machine Algorithm
- Apriori Algorithm
- Linear Regression
- Logistic Regression
- Artificial Neural Networks
- Random Forests
- Decision Trees
- Nearest Neighbors

The models will be built using TensorFlow, Caffe 2.0, and Keras models, with several integration points of data being transformed and delivered with the models. The analytical outcomes will plot the results based on the input data and create graphs. The results typically will include outcomes that will provide identities of customers who are likely to churn. To provide more details on the reasons and the linage of data, the model will use historical data on former churners and will find similarity with existing customers. The similarity will be tagged and identified in the current stack of customers being processed and classifies those customers as potential churners. This classification can be repeated every time for every customer and also groups of customers across geographies. All the data will be managed and computed in the infrastructure.

Methods/techniques used to build the churn model and these include but are not limited to the following:

- Data mining classification techniques
  - Neural networks
    - Performs very well with all data once trained and unleashed for execution
    - Requires familiarity with the model and outcomes to understand the uncovered patterns in the underlying data.
    - Often being thought of as a black box
    - Tend to be relatively slow during learning periods
    - Training the neural network is an exercise that is time consuming but the return on investment is multi-fold
  - Logistic regression models
    - Can give very strong insight into which variables are likely to predict the event outcome
    - To predict an outcome variable that is categorical from predictor variables that are continuous and/or categorical
    - Used because having a categorical outcome variable violates the assumption of linearity in normal regression
    - The only "real" limitation for logistic regression is that the outcome variable must be discrete. Logistic regression deals with this problem by using a logarithmic transformation on the outcome variable which allows us to model a nonlinear association in a linear way.
    - It expresses the linear regression equation in logarithmic terms (called the *logit*)
  - Decision trees
    - Easy to use
    - Shows which fields are the most important
    - Can be vulnerable to noise in the data
    - Leaf in the decision tree could have similar class probabilities

The model can be designed and developed with either logistic regression or a decision tree. The outcome from either use case will identify the most important variables.

The following are application predictions which can be plotted and displayed as outcomes:

- Younger people are more likely to churn and this is true

- Customers who belong to a branch that has been modernized are more likely to churn, this is technology friendly and empowered banks, and this is true

This is one critical application that has been built by banks for using big data platforms, where data can be collected for several users, several dates, several locations, several interactions, several outcomes, and several analyses, leading to conclusive analytics. This is a simplified flow of the actual application development, but the points to understand are as follows:

- You need to become digital
- You need to have ideas that are made to stick
- You need to include neural networks to manage instant analytics
- You need to be proactive with customers
- Existing customers are known entities and can be managed with more accuracy
- Sentiment analytics are important, they are not survey questions however
- Delivering security is essential and important

In conclusion I believe we have reached the tipping point of the *uber* moment in banking and as discussed in this chapter, the industry applications and algorithms are all ready to be leveraged and used as needed.

Additional reading

Big Data, Mining, and Analytics: Components of Strategic Decision Making By Stephan Kudyba.

Disrupting Finance: FinTech and Strategy in the 21st Century edited by Theo Lynn, John G. Mooney, Pierangelo.

Digital Transformation in Financial Services by Claudio Scardovi.

Handbook of Digital Currency: Bitcoin, Innovation, Financial Instruments edited by David LEE Kuo Chuen.

http://www.theguardian.com/business/2015/nov/25/banking-facing-uber-moment-says-former-barclays-boss.

https://www.washingtonpost.com/news/wonk/wp/2016/04/19/say-goodbye-to-your-neighborhood-bank-branch/.

Travel and tourism industry applications and usage

Travel and big data

The global travel industry is expected to grow to 10% of Global GDP by 2022, or an annual revenue of around $ 10 trillion. This industry is at a tipping point where the implementation of prescriptive and predictive analytics, artificial intelligence and neural networks can improve the actions, increase revenue, get more traction with the customers, and overall transform the industry to its next iteration of growth.

In a long running tradition in the industry, travel companies are known for capturing and storing massive amounts of data. During every step of every travel journey they collect data including customer data, flight paths, driving paths, bus journeys, train journeys, transactions, driving styles, check-ins etc. CRM packages implemented at hotels collect vital data from customers for both regulatory compliance and the marketing requirements, and let us not forget yield revenue management was invented in the travel industry already years ago. The data collected sounds very interesting but the business value from this data was difficult to deliver and analyze due to the limitations of the underlying infrastructure. The evolution of computing platforms driven by eCommerce companies and the sheer amount of processing power, cheap, and powerful storage solutions such as Hadoop and many technology vendors waiting to help out, this information can finally be put to use to make the customer feel more appreciated and better serviced, resulting in more revenue and higher profits.

The big question that you want to ask but are unsure of the answers is how do I get value delivered from this data? The answer to this question is to first layout the different blocks of data that you need to use for answering a question whether business value or not. The data blocks will provide you an insight on the interconnectedness of the data; the missing pieces will reveal themselves and provide you opportunity to create a connected data chain. Once this exercise is complete, then you set out to build the application required to answer the question in your mind, in fact my suggestion is to start with a search bar application to see how much of intuitive reply you will get for a search and then start exploring the system. Fig. 8.1 shows the data block layout.

The data block can be built as pyramid layers with increasing order of summarization and aggregation. This layout is essential if you are looking at big data which can be chaotic if not understood well.

Building Big Data Applications. https://doi.org/10.1016/B978-0-12-815746-6.00008-9

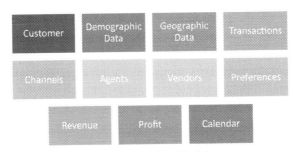

FIGURE 8.1 Sample data block layout.

In a travel industry perspective, the bottom of this data layer is a swamp as it contains raw data and dirty data which are not aligned to the user yet. This is the layer for data discovery and in travel industry the data layer contains searches, links, log times, clickstream data, reprocess, purchases, repeated searches based on price, calendar, destination, links to hotels, and reservations aligned with travel, searches for tourism packages, deals, and incentives. The data can be in multiple languages, have images, videos, user group comments, independent reviews, and sentiment outbursts. This is the world of data which we need to build an application for performing data discovery, data analysis, data segmentation, data classification, and data categorization; data disconnects and provide outcomes on how to connect all this data to deliver value. This is the first application to build.

Once we have streaming data identified as a source, we also see that there are several opportunities missed where the customer or prospect could have received personalized services based on their searches and how that tipping point could have shifted the behavior. Especially in the travel industry, such a personal approach is of vital importance and the opportunities for delivering analytical services at that layer of personalization to each prospect is very important in the travel industry. If we look at the conversion rate on travel websites, we see that approximately 92% of shoppers will not convert and 60% of visitors never return after a first visit. This behavior has been transformed into a positive outcome with engaged stay and in many cases a purchase by the shopper, all of this happening with the utilization of big data and analytics across online travel companies to deliver the right message at the right time to the right person, and provide services which ultimately deliver revenue.

The application storyboard for the operational and raw data analytics including streaming and online data, is that knowing what your customers like, or do not like, can have a severe impact on your brand, its services, and revenue. Everyone shares information on social networks, especially personal travel stories. Collecting and studying this data with a neural network algorithm for text analytics will deliver vital information that can be used to provide a user a more tailor-made message. The goal of the storyboard is to increase stickiness and realize revenue. The best example of this is the "tripadvisor" website.

Luckily, but slowly travel organizations are adapting to big data and more organizations are embracing the idea to integrate business teams and data teams into the data scientists' teams and start experimenting with delivering value. A very interesting use case here is the Mac versus Windows price strategy at Orbitz, where Mac users were steered to 30% higher prices. While it was clear that this was a mistake, it shows the potential big data has for the travel industry. We have seen the increasing potential of kayak as a search engine for travel or CheapOair for deals and packages, these organizations have come to this strength based on their adoption to big data and applications from that platform.

The second level of application to build for use in travel industry is the data lake. This data layer will have all data identified, classified, master data segmented and linked, metadata aligned, ready for aggregation, analytics, and in-depth analysis. The applications in this layer include Tableau mash-ups, R and SAS models, TensorFlow and Caffe networks, and Solr and Elastic search engines for analyzing the data.

The third level of applications to build is the neural network and machine learning-driven applications which will deliver value to both the prospect in the travel search and booking, and the internal teams analyzing data to deliver business value. The world we live in today is driven by these algorithms and applications, and we are expecting more value from every bit of data we generate. The world is now called as "Internet of Things". Our applications are today built for the cloud, delivered as containers and can be driven to align with our expectations as the machine learns you quickly, but misses the critical point of emotions. However we can drive the decision based on a decision tree algorithm along with others and provide steps that can be altered as needed based on user reactions and add more validation to the logic as the process evolves.

The challenge is to connect all the data across the different platforms, websites, or products during the journey of a traveler. A wish-list item in early 2000s was to have a traveler informed along the journey with messages if the plane is delayed, including the new gate. The flight delays and associated plan changes are also intimated to the hotel that is booked, that would truly be exceeding expectations. Today we are doing all of these and more. How did we get here, who delivers the innovation?

The inside innovation is driven by the thought leaders of the industry along with the data architects, statisticians and analytics experts. These teams are aided with infrastructure and have access to all the data and the business subject matter experts. However there is an outside innovator too? Who? Read on.

As a traveler, I have always wondered how we get so many pushed notifications and alerts, in a continuum and throughout the journey. The interesting aspect to understand is we the customers are the outside innovators to the travel enterprises. We use our smartphones to connect and have our boarding passes, once that is done we are ready to provide continuous information on our move, the airline reacts with push notifications. Along the way you get updates on flight time, if you are on an upgrade list and are getting an upgrade, you also start connecting the landing activities including rental cars or uber, hotel messages for alerting you on check-in, and then there are transactions that happen along the way that your credit card company or bank can harvest ideation of travel. These data points were somewhat available prior to the Internet, but the emergence of the eCommerce platform transformed all the aspects of data and the evolution of neural networks since 2010 has delivered very impactful decisions and insights.

The impact of big data applications has been seen across airports globally. In the United States, we have seen major transformations in the San Francisco International Airport, Dallas Fort Worth International Airport, Chicago Midway International Airport, and many others. These changes that were implemented were done by collecting twitter posts and social media posts on how and what travelers felt will improve their flying experience across the airport. For example, in San Francisco airport there were no rooms for new mothers or for yoga; many travelers had indicated these are essential for an airport. The authorities took the messages and ran several online polls to validate their findings and eventually we saw the transformation happen. Today there are so many positive tweets and feedback about the improvements and the outcomes have driven revenue to a higher margin for all vendors and airlines in the airport. In a recent conference, an analyst from Schiphol Airport, explained that big data is also everywhere at Schiphol airport. Big data is used to measure the amount of people present at Schiphol in real time, to develop heatmaps for expected noise pollution in the surroundings and to visualize retail sales versus departure gates to see how far travelers wander off from their departure gate. I have personally experienced this having traveled through this airport at least two to three times in a year. These are just a few examples, and we can see the expanding potential of big data in the passenger journey is tremendous. Today in the world driven by Internet, speed is everything. Consumers generally move away within seconds if an online answer takes too long. After all there are many other websites around offering exactly the same. Each of these websites need to sift through millions of records from various sources such as airline agencies or global distribution companies and delivering a result earlier can have a direct impact on revenue. So, speed matters and when speed matters, big data is the answer. By building their own big data system for example, a German travel company is now processing over 15,000 queries per second while searching through 26 billion offers across 35 parameters and deliver an answer within the second. This is another application of data in the space to deliver the right results to the users within the expected time.

Implementing the combination of neural network algorithms and machine learning is the booking company Hopper. With Hopper it will be possible to plan and book a trip just by entering a vague idea that you have in their search engine and the algorithms will do the rest. It might be very difficult to combine all that unstructured data that is out there to deliver the best result and experience to the guest, but it is the only way forward. Travel organizations, if they have not already done so, should wake up and start analyzing the right big data.

As seen from the discussions in the travel segment, there are several opportunities and globally we are on a huge platform that will provide all players with benefits if they harvest the right moments. Here are some of the examples and benefits:

Real-time conversion optimization

By leveraging internal and external data, hospitality businesses can achieve razor sharp targeting, reduce customer acquisition costs, and increase customer lifetime value. The best part—all of this can happen in nearly real time.

According to Skift, companies' leveraging traveler data tends to:

- Reduce customer acquisition costs by 21%.
- Generate a 17% increase in hotel/vehicle reservations.

By pairing internal marketing data with external sources, hotels can improve operations even further. The simplest example would be leveraging weather data to adjust the current offerings and estimate the possible bookings. A skiing-based property can proactively outreach to customers with personalized offers whenever more snow is expected. Or, on the contrary, pitch additional leisure activities and offers to guests when not much powder is available on the slopes.

More analytics with data will lead you to capitalize on micromoments happening in nearly real time. A popular midtier hotel chain estimated that around 90,000 passengers in the United States were stranded every day due to flight cancellations. Their marketing teams developed an algorithm to track flight disruptions in real time and trigger targeted mobile search ads for nearby property locations. This "micromoment" based campaign generated a 60% increase in bookings.

Optimized disruption management

Today weather and disruptions caused by the same to travelers is an everyday and multiple times in a day occurrence. In this realm of affairs, data science is a powerful tool to deliver instant help and response to affected travelers. The algorithms can be trained to monitor and predict travel disruptions based on the information at hand—weather, airport service data, and on-ground events such as strikes and so on.

The application will be executed $24 \times 7 \times 365$ and the system will be trained to alert travelers, staff, and management about the possible disruptions and to create a contingency action plan in response.

The same application can be executed at the travel agency, and it will setup chatbots and automatically assign personal assistants who will be able to assist affected travelers and help them adjust their travel plan. According to a PSFK travel debrief survey, 83% of industry experts say that control over their own travel experience through real-time assistance will be very important for travelers. Airlines can also ramp up their disruption management plans with the help of data analytics.

A leading Australian airplane company was among the early adopters of a third-party schedule recovery system, driven by neural networks and machine learning algorithms. The system was used to report inclement weather and predict potential cancellation of flights. The system once put into production helped them reduce the number of delayed/canceled flights due to bad weather conditions. Their competitor, who had used a manual system to manage disruptions, canceled 22% of flights, whereas this company reduced the number to 3.4%. This is a real life application of big data analytics and applications.

Niche targeting and unique selling propositions

Data science is not reserved for the big brands only. Smaller travel companies can leverage their existing data sets to become the best within their segments, instead of competing for every/any kind of customer.

A boutique hotel chain has properties in multiple locations. They collect customer data from travelers and have a lot of valid information that can be leveraged to execute several analytical algorithms that will deliver outcomes to hone their unique value proposition. Instead of pursuing a large pool of target clients, the chain's CDO decided to focus on pursuing a microlevel niche demographic, travelers with a specific business-level budget.

To execute the nearest neighbor algorithm, the hotel chain aggregates on-site data into a central dashboard system that allows managers at all locations to review every guest interaction, and obtain further insights on how to improve customer experience. The company specifically focuses on attracting the one particular type of customer and builds strong matches between customers and properties. The accumulated data is used to make better decisions about services and craft more targeted marketing campaigns targeting "look-alike" prospects. Small businesses in the travel industry can follow the botique's lead and maximize their internal data to pursue the right customer, instead of wasting budgets on ineffective marketing to a broader segment.

"Smart" social media listening and sentiment analysis

Social media is a two-way communication alley. Sharing updates are not enough to succeed. You have to communicate and listen to your customers and prospects. Applying data science to social listening can help marketers consolidate large amounts of scattered data and turn it into specific market research campaigns. Most travel companies have accounts registered on popular networks, yet they view "social media" as a separate entity, rather than an organic extension for their marketing business.

Travel brands can leverage social media to create highly targeted buyer personas and scout for "look-alike" prospects. All the conversations happening around the customer or prospect can be collected and studied, analyzed, and refined to the key points that are being discussed. The new data can then be used in machine learning execution of unsupervised data, where the patterns are more identified and can be leveraged to attract potential customers.

Hospitality industry and big data

The term "big data" means many different things to different people. Generally speaking, it describes the collection of information from both offline and online sources to support the ongoing development and operations of a company or organization. This segment of discussion is around the hospitality industry which is related to hotels, resorts, and vacation properties.

Data can originate from almost anywhere, including everything from historical records, point-of-sale devices and current consumer habits to customer feedback, referrals, and online reviews. With so much information coming from all directions, it is tough to rein it all in and apply it a useful way, but that is where the discipline of big data analytics comes into play.

It is crucial to define big data when trying to understand its role in hotel and hospitality. With so many companies embracing big data and applying it to improve their operations, it might be the key to maintaining competitiveness in the future. Big data has relevant applications in nearly every industry imaginable. To use this information to its fullest extent, it is important to know where it has the greatest impact.

- Customer categorization is a critical success factor for maximizing profits regardless of industry. This algorithm is used in the categorization and identification of incoming customers. To properly categorize a customer according to their potential for profitability, companies must track customer spending habits over time, track their channels of prospecting, their influencers, and understand their social makeup and behavior. These pieces when aligned together in a model will help develop analytics and provide the appropriate bucket to segment the customers for campaigns. A leading financial services institution was using incorrect segmentation models leading to 2% return on campaigns, when this was noticed and corrected with all associated inputs, the same campaign execution resulted in 34% return and an additional 21% cross-selling opportunities. Hotels can leverage similar approaches to customer segmentation and award them buckets of points and levels of segmentation, which will promote their loyalty and keep them as happy customers. A happy customer is the best referral that can be available for more growth and visibility.
- Personalized service is an opportunity to hotels to provide customers with benefits that make them feel rewarded. For example if you travel 14 h by flight and then spend 2 h at customs before you arrive at a hotel, then best personalized services to offer you would be easy check-in, a good breakfast, a massage service for 1 day, and offer to assist you in any travel or other services. This type of a service exists in hotels, but not for all customers and this can be easily modified to accommodate all types of customers, except that some customers might have to pay for certain services, which can be compensated in charges. Today hotels are offering

customized services to cater to their most valuable customers—either in groups or as individuals. But hotels cannot always rely on internal data to predict a customer's return. Instead, data analysts have to collect data from surrounding, external sources to help identify travel patterns, habits, and common timeframes, to make this prediction. Nevertheless, the customer feels really appreciated from personalized service. There can be a survey that can be done with simple yes and no answers to receive feedback on the services and further tweak and improvements can be offered in subsequent stays and visits.

- Social media—In today's Internet-centric age, online communities are just as important as their traditional counterparts. Since many customers turn to social media for questions, issues, and concerns, the platform provides a great opportunity to connect with consumers in brand-new ways. Several airlines in the Middle East have compiled vast datasets containing online search histories, completed bookings, and even airport lounge activities on every one of their customers more than 200 million of them. The data helps officials create personalized travel experiences for their frequent guests. The frequent guests in-turn have brought more fliers to enroll into activities which have driven revenue for the airlines across different airports.
- Yield management—Big data analytics also affects yield management. By calculating the optimum value of each room and factoring in metrics like seasonal demands, regular guests, and similar trends, hotels can ensure maximum profits. There are several algorithms that can be used to run these calculations.

Regardless of whether hotels are trying to classify their patrons with better accuracy, provide personalized services, engage their social media audience, or stretch the value of their properties; they must use and apply all this data before it has an impact. The information on its own is dormant until activated through the disciplines of big data processing and analysis.

Analytics and travel industry

The analytics that we used in the industry was limited in the value it provided as the infrastructure was limited in storage and compute. With the big data platforms and the issue of storage and compute being resolved the analytics are becoming more relevant and can be utilized for multiple purposes. The analytics that we will use the most include the following:

- Descriptive analytics is used to analyze data from past occurrences and activities, used commonly by marketing and advertising. Predictive analytics uses big data to try and forecast future outcomes or events, while prescriptive analytics takes advantage of highly advanced algorithms to process big data and provide actionable advice. All three of these methods are common strategies for applying big data in the hotel and hospitality industry.

- Descriptive analytics—This method is one of the most straightforward and efficient ways of generating actionable data. Did a recent renovation increase sales, or was it ultimately a waste of capital? It is easy to answer questions like this via descriptive analytics—it is a decades-old method that has assumed many different forms over the years.
- Predictive analytics—Basic examples of predictive analytics include preparing a hotel for a seasonal rush—like spring break—and reducing the hours of staff members to accommodate the fewer number of reservations in the offseason.
- Prescriptive analytics—Instead of letting the human workforce interpret and act on this information without any guidance, some of today's systems provide recommendations and advice to improve service and increase profits. Online reservation systems that track a guest's past stays can automatically generate discount codes for future reservations, assemble personalized services for each guest, and even deliver their favorite drinks or food.

Big data analytics has the potential to completely transform the customer experience within the hotel and hospitality industry. It is not something that will happen overnight, but the industry is already making huge strides toward a full-on embrace of big data and all the advantages it has to offer.

Some of the most tech-savvy hotel chains are already adopting long-term strategies and policies for big data management. Those who are unwilling or hesitant might find it hard to compete in the coming years. The niche of big data is still in its infancy, but it is already sparked storms of creativity and innovation in any industry it has touched, including hotels and hospitality. Even the most sophisticated of predictive analytics cannot tell us exactly where big data is headed, but customers are sure to be pleased with the results.

Examples of the use of predictive analytics

Recommender systems for travel products (e.g., hotels, flights, ancillary services)—There are thousands of possible combinations of flights connecting Los Angeles and New York for example, and this figure breaks the roof when combining possible services. But which travel solutions and services are relevant for a given passenger? Which hotel is the most pertinent for a young couple who just booked their flights for next summer holidays? Recommender systems provide win–win value for both users and travel providers by proposing the most valuable and relevant options to users while maximizing revenues of travel providers. Predictive analytics help to better understand user needs and match this knowledge to possible products and services.

Video, image, and voice recognition systems for travel purposes—Our human brains respond to stimulus coming from different senses. They are better adapted to understand natural forms of communication-like images of sounds rather than textual written information. With the development of deep learning and other AI algorithms, the processing of this unstructured data is not *Sci-Fi* anymore. Machines are now able to understand images and sounds, and in some cases, even better than the human brains. This brings new opportunities for applications in the travel industry: from inspiration (where to go?) to automation of reservations.

Click and conversion optimization for travel products and online advertising campaigns— Online marketing is all about conversion; that is, the ability to sell products or services with minimum exposure. Attracting users to an ad is not enough if customers are not buying. New predictive algorithms could estimate conversion and help better define travel products, better place the ads, and finally optimize advertising campaigns. These advanced algorithms, fed with enriched travel data, become more and more able to understand passenger needs, notably based on a combination of hundreds of specific factors that have been found to be relevant for travelers. This opens new areas and unlocks huge opportunities for online advertising.

Social media analysis (e.g., sentiment analysis and profiling)—Monitoring social networks is a strategic task and it is not possible to do it manually anymore. For example, it has been estimated that 90% of American travelers (who have a smartphones) share photos and experiences about their travels on social networks. Similarly, millions of travel-related reviews are shared on the internet every day. Sentiment analysis permits the estimation of the polarity of these posts (e.g., reviews, tweets) in milliseconds. That means knowing if they are positive, negative, and so forth. Predictive analytics have been also been used on social networks to better known users, their interest, and needs. Just tell me who your friends are and how you write and the algorithms will tell you who you are (or what you look like).

Alerting and monitoring—The travel industry generates huge volume of data. For example Amadeus process more than 1 billion transactions per day in one of its data centers. New aircraft have close to 6000 sensors generating more than 2 Tb per day. Obviously this data cannot be analyzed by human beings. Using supervised machine learning algorithms, known defects can be anticipated when a combination of factors is observed much like how a set of symptoms helps doctors diagnose a particular disease (with some probability). On the other hand, unsupervised learning algorithms have helped detect anomalies to generate alerts when some data observation becomes suspiciously rare.

Develop applications using data and agile API

Flight APIs—The flight APIs allow you to find prices of flights from any given origin to any destination. One of the APIs does not even require a destination, while giving you the best prices for a variety of cities. It is ideal for inspiration applications aimed at offering several options to the traveler. The flight APIs also provide airport information. You can find the closest airport to a given location and even help the traveler autocomplete destination or origin forms in your application prototype.

Hotel Shopping APIs—Similar to the flight APIs, the hotel shopping APIs allow you to return availability and price for accommodation at any given location. You can either search using an IATA airport or city code or actually define the search zone yourself. These APIs also allow you to get detailed information on any hotel, including address, amenities, awards, room, prices etc.

Cars Shopping APIs—The car shopping APIs provide you with a comprehensive way to search for a rental car. You can either search using IATA airport codes or define a geographical zone, making it the perfect way to enrich the traveler's booking flow on your prototype.

Rail and **Train APIs**—The rail and train APIs allow you to find schedules, availability, and prices for trains in France at any given date. Adding train to your flight and hotel booking application is a great way to create a fully integrated journey for the traveler.

Travel Record API—The travel record API returns the trip information for a given passenger name and reservation code. This API can do wonders when it comes to creating personal traveling assistants to guide passengers during their trip.

Travel Intelligence APIs—The new travel intelligence APIs allow you to access travel insights based on real data. They allow you to find out the most popular destinations people are searching and traveling to from a given origin. Similarly, the travel intelligence APIs allow you to check the traffic between two airports over time.

In conclusion we see a huge upliftment of the customer and their 360 degrees of activity becoming the focal point of the travel industry. Today we see the availability of data and its trends with the movement of the customer, who can be a new or an existing or a returning prospect providing us with perspectives on how they will and shall react, which will be useful to create storyboards and be very useful in prescriptive analytics for the enterprise. We have several innovative solutions deployed across many of the travel and tourism channels in the eCommerce, which have delivered efficiencies and increased revenue cycles by multiple percentages to enterprises.

Additional reading

Mobility Patterns, Big Data and Transport Analytics: Tools and Applications for Modeling by Constantinos Antoniou, Loukas Dimitriou, Francisco Pereira.

Big Data and Innovation in Tourism, Travel, and Hospitality: Managerial edited by Marianna Sigala, Roya Rahimi, Mike Thelwall.

Learning Big Data Gathering to Predict Travel Industry Consumer Behavior Johnny Ch Lok.

Research Methods for Leisure and Tourism by A.J. Veal.

The Analytic Hospitality Executive: Implementing Data Analytics in Hotels by Kelly A. McGuire.

9

Governance

Definition

Governance is the process of creating, managing, and delivering programs which always focus on the corporate goal of delivering business value from data. This goal is a dream come true when you can successfully orchestrate the program from gathering all the data to harnessing wonderful and valuable business insights that increase the revenue for the enterprise and deliver value to the customer.

Data is the biggest assets of an enterprise. With the right set of information, making decisions regarding the business always provides the right levels of confidence as you can audit and attribute the data you used for the decision-making process. The complexities of managing applications which need to consume both structured and big data ecosystems are not a trivial task. Enterprises that have been through this journey will tell you. It is an ease of having governance as a part of the organization.

Data management refers to the process of collecting, processing, storing, and distribution of data. The data management techniques that we have been using today in the modern database world, is based on requirements that were developed for legacy systems dating back from punch cards to mainframes to the analytical data processing systems.

Fig. 9.1 shows the fundamental stages of data management across enterprises. There are several cycles of activities within each stage of processing that creates complexities of managing the entire process end to end. Before we look into the details of these processes, let us take a brief look at metadata and master data.

Metadata and master data

Metadata is defined as data about data or in other words information about data within any data environment. The origins of metadata can be traced to library systems from many years ago, where the classification, tagging, and catalog of books provided the fundamental information classification and retrieval. Applied to information technology, metadata provides a natural way to integrate data and interrogate data.

Why is metadata so valuable? The answer to this question can be explained by a simple example—let us assume you make a trip to the local hardware store to buy some sheets of wood. Without any label describing the wood, its dimensional attributes and price, you will be lost for hours waiting for someone to help you. This little label of information is what the value of metadata is all about.

Building Big Data Applications. https://doi.org/10.1016/B978-0-12-815746-6.00009-0

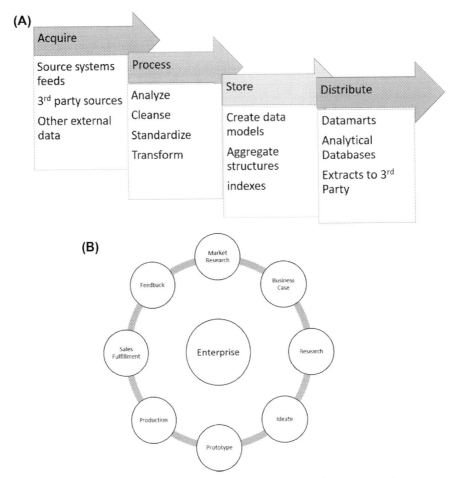

FIGURE 9.1 (A) Data management stages. (B) Sixth sense advisors perspective.

From a database perspective, metadata describes the content of every attribute in every table as defined during the design phase of the database. This provides the developers, data architects, business intelligence architects, analysts and users a concise roadmap of what data is stored within the database and in which table.

Metadata changes in the lifetime of a database, when changes occur within the business such as mergers and acquisitions, new system deployment, and integration between legacy and new applications. To maintain the metadata associated with the data, we need to implement business and technology processes and governance policies. Many enterprises today do not track the lifecycle of metadata, which will cost them when data is brought back from backups or data is restored from an archive database, and nobody can quite decipher the contents and its relationships, hierarchies, and business processing rules.

The value of having metadata can be easily established in these situations by measuring the cost impact with and without metadata:

- Cost of commissioning new applications
- Learning curve for new employees
- Troubleshooting application problems
- Creating new business intelligence and analytics applications
- Data auditing
- Compliance auditing

Traditionally in the world of data management, metadata has been often ignored and implemented as a postimplementation process. When you start looking at big data, you need to create a strong metadata library, as you will be having no idea about the content of the data format that you need to process. Remember in the big data world, we ingest and process data, then tag it, and post these steps consume it for processing.

Let us revisit the metadata subject area first and then understand how it integrates into the world of big data.

Source of metadata include the following:

- Metadata generated automatically for data and business rules
- Metadata created by the designer of the systems
- Metadata procured from third party sources

There are fundamentally five types of metadata that are useful for information technology and data management across the enterprise from transaction processing to analytical and reporting platforms, these include the following:

- Technical metadata—consists of metadata that is associated with data transformation rules, data storage structures, semantic layers, and interface layers.
  - Metadata for data model and physical database includes length of a field, the shape of a data structure, the name of a table, the physical characteristics of a field, the number of bytes in a table, the indexes on a table, and DDL for the table
  - Business processing metadata includes information about:
    - The system of record for a specific piece of data,
    - Transformations that were performed on what source data to produce data in the data warehouse/data mart,
    - Tables and columns used in the particular process in the data warehouse/data mart and what do the transformations mean
  - Administrative metadata.
- Business metadata—Business metadata refers to the data describing the content available in the data warehouse/data mart and describes the following:
  - The structure of the data
  - The values that are stored within the attributes

- Create and update date and time
- Processing specifics if any for the field or attribute
- Contextual metadata—contextual metadata is data that sets "context" of your data. This is more related to processing large objects like text, images, and videos. Examples of contextual metadata include news feeds, tags on images etc. Contextual metadata is the hardest to collect.
- Process design level metadata
  - Source table
  - Target table
  - Algorithms
  - Business rules
  - Exception
  - Lookup or master data or reference data
- Program level metadata
  - History of transformation changes
  - Business rules
  - Source program
  - System name
  - Source program author
  - Extract program name and version
  - Load scripts
  - Script specific business rules
  - Load frequency
  - Extract dependencies
  - Transformation dependencies
  - Load dependencies
  - Load completion date/time stamp
  - Load completion record count
  - Load status
- Infrastructure metadata
  - Source system platform
  - Source system network address
  - Source system support contact
  - Target system platform
  - Target system network address
  - Target system support contact
  - Estimated size (tables/objects)
- Core business metadata
  - Field/object description
  - Confidence level
  - Frequency of update
  - Basic business metadata

- ⊛ Source system name
- ⊛ Valid entries (i.e., "There are 50 States, Select One")
- ⊛ Formats (i.e., RegistrationDate: 20-JAN-2019 18:00:00)
- ⊛ Business rules used to calculate or derive the data
- ⊛ Changes in business rules over time
- ⊛ Additional metadata
  - Data owner
  - Data owner contact information
  - Typical uses
- ⊛ Level of summarization
- ⊛ Related fields/objects
- ⊛ Existing queries/reports using this field/object
- Operational metadata
  - ⊛ Information about application runs:
    - Frequency
    - Record counts
    - Usage
    - Processing time
    - Security
- Business Intelligence Metadata—contains information about how data is queried, filtered, analyzed, and displayed in business intelligence and analytics software tools
  - ⊛ Data mining metadata: The descriptions and structures of datasets, algorithms, queries
  - ⊛ OLAP metadata: The descriptions and structures of dimensions, cubes, measures (metrics), hierarchies, levels, drill paths
  - ⊛ Reporting metadata: The descriptions and structures of reports, charts, queries, datasets, filters, variables, expressions
  - ⊛ Business intelligence metadata can be combined with other metadata to create a strong auditing and traceability matrix for data compliance management

Metadata is very essential to manage the lifecycle of data from an enterprise perspective. We have landed ourselves into this infinite loop of processes and projects due to the absence of streamlined, governed, and managed metadata. This aspect is a big impact issue to be addressed for building big data applications. One of the most successful enterprises in managing metadata is Procter&Gamble, who can provide you research formula based on your question specific to geography, water, soil, and environment conditions. How did they do this? When the labs started 50 years ago, they had to collect and maintain all experimental data; this included the experiments both in the labs and during stay at home and listen to consumer experience, thus came the need to add metadata which has grown into corporate culture. How cool and easy it is to walk through zettabytes of data all of them tagged and organized. This is the key lesson to learn from successful corporate cultures. Our next segment focuses on another important member of the data family, called as the master data.

Master data

Master data management is used by applications to add key business entities (customer, products, policy, agent, location, and employee) while processing and utilizing data from the data layers within an organization. The master data is processed and stored in a master data management database (MDM database).

Why is master data management important? In the traditional world of data and information management, we used to build data and applications in silos across the enterprise. The addition of new systems and applications resulted in not only data volumes and transactions, but also created redundant copies of data, and in many cases the same data structure contained disparate values. The end state architecture resulted in systems that did not interface and integrate with each other. The complexity of processing disparate data into a common reference architecture required hours of manual effort and did not provide a clean and auditable result set. Each system can give you a fractured insight into what is happening in the enterprise but you cannot create a clear and concise view of data as a centralized system.

This is where the need for a centralized master data management system begins. With a centralized system, the enterprise can create, manage, and share information between systems and applications seamlessly. The efforts to manage and maintain such a system is very simple and flexible compared to a decentralized platform. This approach can save the enterprise time and opportunity costs, while ensuring data quality and consistency. Master data management is driven to handle each subject area as its own system within the enterprise. The underlying architecture of the system allows multiple source systems to be integrated and each system can alter the attribute values for the subject area. The final approval of the most accurate value is determined by a Data Steward and a Data Governance team, post which the business rules are executed to process the data modifications. The results are then shared back with the source systems, applications, and downstream analytical applications and databases, and called as the "gold copy" of the data definition.

Master data management is not about technology, the critical success factors this initiative are the subject matter experts in data within the business teams, who can understand and define the processing rules and complex decision-making process regarding the content and accuracy of the data. Master data management is not implementing a technology, as the role of any technology platform in this process is that of a facilitator and an enabler.

Master data management is about defining business processes and rules for managing common data within disparate systems in the enterprise. In implementing these processes, the data governance and stewardship teams collectively determine the policies, validation and data quality rules, and service-level agreements for creating and managing master data in the enterprise. These include the following:

- Standardized definition of data common to all the systems and applications
- Standardized definition of metadata.
- Standardized definition of processes and rules for managing data
- Standardized processes to escalate, prioritize, and resolve data processing issues.
- Standardized process for acquiring, consolidating, quality processing, aggregating, persisting, and distributing data across an enterprise.
- Standardized interface management process for data exchange across the enterprise internally and externally
- Standardized data security processes
- Ensuring consistency and control in the ongoing maintenance and application use of this information

Metadata about master data is a key attribute that is implemented in every style of master data implementation. This helps resolve the business rules and processing conflicts that are encountered by teams within organizations and help the data governance process manage the conflicts and resolve them in an agile approach.

Data management in big data infrastructure

With the world of big data there is a lot of ambiguity and uncertainty with data that makes it complex to process, transform, and navigate. To make this processing simple and agile, a data-driven architecture needs to be designed and implemented. This architecture will be the blueprint of how business will explore the data in the big data side and what they can possibly integrate with data within the RDBMS which will evolve to become the analytical data warehouse. Data-driven architecture is not a new concept, it has been used in business decision-making for ages, except for a fact that all the touchpoint's of data we are talking about in the current state are present in multiple silos of infrastructure and not connected in any visualization, analytic, or reporting activity today.

Fig. 9.2 shows the data touchpoint's in an enterprise prior to the big data wave. For each cycle of product and service from ideation to fulfillment and feedback, data was created in the respective system and processed continuously. The flow of data is more of a factory model of information processing. There are data elements that are common across all the different business processes, which have evolved into the masterdata for the enterprise, and then there is the rest of the data that needs to be analyzed for usage, where the presence of metadata will be very helpful and accelerates the data investigation and analysis. The downside of the process seen in Fig. 9.1 is the isolation of each layer of the system resulting in duplication of data and incorrect attribution of the data across the different systems.

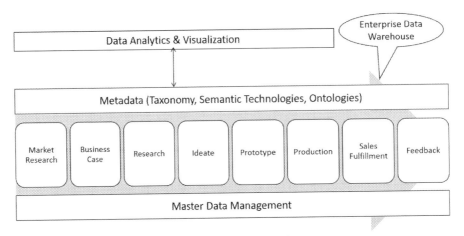

FIGURE 9.2 Enterprise use of data pre big data.

The situation shown in Fig. 9.1 continues to happen with the best data governance programs implemented due to the fact that organizations continue to ignore the importance of corporate metadata and pay the penalty once incorrect information is processed into the systems from source systems all the way to business intelligence platforms.

Fig. 9.3 shows the data-driven architecture that can be deployed based on the metadata and master data solutions. This approach streamlines the data assets across the enterprise data warehouse and enables seamless integration with metadata and master data for data management in the data warehouse. While this architecture is more difficult to implement, it is a reusable approach where new data can be easily added into the infrastructure as the system is driven by data-driven architecture. Extending this concept to new systems including big data is more feasible as an approach. Let us take a quick look at processing traditional data with metadata and master data before we dive into applying this approach to processing big data and enabling the next generation data warehouse to be more data driven and agile.

FIGURE 9.3 Enterprise data-driven architecture.

Fig. 9.4 shows the detailed processing of data across the different stages from source systems to the data warehouse and downstream systems. When implemented with metadata and master data integration the stages become self-contained and we can manage the complexities of each stage within that stage's scope of processing, as discussed next:

- Acquire stage—In this stage of data processing, we simply collect data from multiple sources and this acquisition process can be implemented as direct extract from a database to data being sent as flat files or simply available as a web service for extraction and processing.
  - Metadata at this stage will include the control file (if provided), the extract file name, size, and source system identification. All of this data can be collected as a part of the audit process.
  - Master data at this stage has no role as it relates more to the content of the data extracts in the processing stage.
- Process Stage—In this stage of processing the data transformation and standardization including applying data quality rules is completed and the data is prepared for the loading into the data warehouse or data mart or analytical database. In this exercise both metadata and master data play very key roles.
  - Metadata is used in the data structures, rules, and data quality processing.
  - Master data is used for processing and standardizing the key business entities.
  - Metadata is used to process audit data.

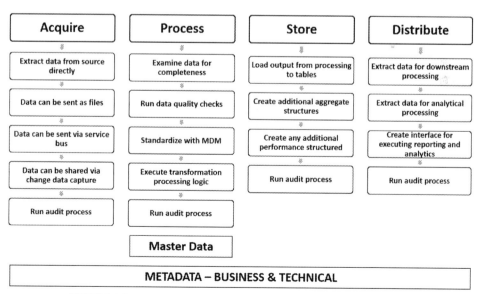

FIGURE 9.4 Data processing cycles with integration of MDM and metadata.

In this processing stage of data movement and management, metadata is very essential to ensure auditability and traceability of data and process.

- Storage Stage—in this process the data transformed to final storage at rest is loaded to the data structures. Metadata can be useful in creating agile processes to load and store data in a scalable and flexible architecture.
    * Metadata used in this stage includes loading process, data structures, audit process, and exception processing.
- Distribution Stage—in this stage, data is extracted or processed for use in downstream systems. Metadata is very useful in determining the different extract programs, the interfaces between the data warehouse or data mart and the downstream applications and auditing data usage and user activity.

In a very efficiently designed system as described in Fig. 11.4 we can create an extremely scalable and powerful data processing architecture based on metadata and master data. The challenge in this situation is the processing complexity and how the architecture and design of the data management platform can be compartmentalized to isolate the complexities to each stage within its own layer of integration. Modern data architecture design will create the need for this approach to process and manage the lifecycle of data in any organization.

We have discussed the use of metadata and master data in creating an extremely agile and scalable solution for processing data with applications in the modern data warehouse. The next section will focus on implementing governance and leveraging benefits from the same.

Processing complexity of big data

The most complicated step in processing big data lies with the not just the volume or velocity of the data but also its

- Variety of formats—data can be presented for processing as excel spreadsheets, word documents, pdf files, OCR data, email, data from content management platforms, data from legacy applications, and data from web applications. Sometimes it may be variations of the same data over many time periods where the metadata changed significantly.
- Ambiguity of data—can arise from simple issues like naming conventions to similar column names of different types of data to same column storing different types of data. A lack of metadata and taxonomies can create a significant delay in processing this data.
- Abstracted layers of hierarchy—the most complex area in big data processing are the hidden layers of hierarchy. Data contained in textual, semistructured, image

and video, and converted documents from audio conversations all have context and without appropriate contextualization the associated hierarchy cannot be processed. Incorrect hierarchy attribution will result in datasets that may not be relevant.

- Lack of metadata—there is no metadata within the documents or files containing big data. While this is not unusual, it poses challenges when attributing the metadata to the data during processing. The use of taxonomies and semantic libraries will be useful in flagging the data and subsequently processing it.

Processing limitations

- Write Once Model—with big data there is no update processing logic due to the intrinsic nature of the data that is being processed. Data with changes will be processed as new data.
- Data fracturing—due to the intrinsic storage design, data can be fractured across the big data infrastructure. Processing logic needs to understand the appropriate metadata schema used in loading the data. If this match is missed then errors could creep into processing the data.

Big data processing can have combinations of these limitations and complexities, which will need to be accommodated in the processing of the data. The next section discusses the steps in processing big data.

Governance model for building an application

The subject is very easy to state but extremely complex when discussed in layers, we will look at the same. Fig. 9.5 is a description of the layers of governance related to the build of applications.

As seen in Fig. 9.5, there are several layers of governance applied in big data applications. The first layer is the data layer, which covers the data management of big data that will be used in the entire exercise, this includes data acquisition, data discovery, daya analysis, data taging. Metadata processing, master data integration, and data delivery to data lakes and data hubs. The complexity of governance is these layers is very essential to understand, the reason being that we will have data discovery, exploration, and analysis being done by many business users across different teams and their outcomes, logs, and analysis need to be sorted, processed, and managed efficiently. We will be delivering these processes and services both as microservices and as robotic process automation exercises, these need to be documented, captured, and stored for easy access and use, the functionality of this exercise is also falling under governance.

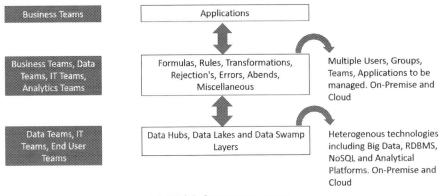

FIGURE 9.5 Governance process.

The formulas, transformation rules, and all associated transformation of data within the data layers and further in the application layer needs governance. This aspect is very critical especially in large scale research experiments like CERN or cancer treatment and research applications. The formulas will need to be tagged with each application it is used by, if it is a library there has to be metadata tags of all applications using it and transforming data. The pivotal issue here is the maintenance of the formula libraries, they need data stewards who know what additions, changes, and deletions are being done, as the teams that consume these libraries are varied and any change can cause unforeseen results, which will wreak havoc. One lesson in this governance strategy is the maintenance of history and version control to be managed by applications and its consumers. The ability to fork a new version allows you to manage the data transformation without impacting the larger team, very similar to what we do with Github. This will provide benefits and increase efficiencies within the team. The rules, transformations, calculations, and all associated data-related operations performed within the application need to be governed by this aspect and it will ensure valid processing of data by each application.

Use cases of governance

Machine learning

From the prior discussions we see that processing big data in a data-driven architecture with semantic libraries and metadata provide knowledge discovery and pattern-based processing techniques where the user has the ability to reprocess the data multiple times using different patterns or in other words process the same dataset for multiple contexts. The limitation of this technique is that beyond textual data its applicability is not possible. At this stage is where we bring in machine learning techniques to process data such as images, videos, graphical information, sensor data, and any other type of data where patterns are easily discernible.

Machine learning can be defined as a knowledge discovery and enrichment process where the machine represented by algorithms mimic human or animal learning techniques and behaviors from a thinking and response perspective. The biggest advantage of incorporating machine learning techniques is the automation aspect of enriching the knowledge base with self-learning techniques with minimal human intervention in the process.

Machine learning is based on a set of algorithms that can process a wide variety of data that normally is difficult to process by hand. These algorithms include the following:

- Decision tree learning
- Neural networks
- Naive Bayes
- Clustering algorithms
- Genetic algorithms
- Learning algorithms
  - Explanation-based learning
  - Instance-based learning
  - Reinforcement-based learning
  - Support vector machines
- Associative Rules
- Recommender algorithms

The implementation of the algorithms is shown in Fig. 9.6. The overall steps in implementing any machine learning process are as follows:

1. Gather data from inputs
2. Process data through the knowledge-based learning algorithms, which observes the data patterns and flags them for process. The knowledge learning uses data from

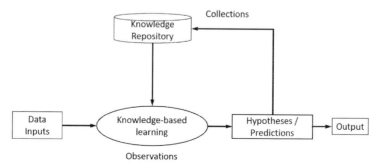

FIGURE 9.6 - Machine learning process.

prior processing stored in a knowledge repository (a NoSQL or DBMS-like database) along with the algorithms for machine learning.

3. The data is then processed through the hypothesis workflows
4. The output from a hypothesis and predictive mining exercises are sent to the knowledge repository as a collection with meta-tags for search criteria and associated user geographic and demographic information as much available.
5. Process the outputs of hypothesis to outputs for further analysis or presentation to users.

Examples of real-life implementations of machine learning:

- IBM Watson
- Amazon recommendation engine
- Yelp ratings
- Analysis of astronomical data
- Human speech recognition
- Stream analytics
  - Credit card fraud
  - Electronic trading fraud
- Google robot–driven vehicles
- Predict stock rates
- Genome classification

Using semantic libraries, metadata and master data along with the data collected from each iterative processing, enriches the capabilities of the algorithms to detect better patterns and predict better outcomes.

FIGURE 9.7.1 User searches for movie.

Let's see how a recommendation engine uses all the types of data to create powerful

FIGURE 9.7.2 User gets result sets.

and personalized recommendations. We will use Amazon website to discuss this process (Figs. 9.7.1, 9.7.2 and 9.7.3)

1. John Doe searches for movies on Amazon.
2. John Doe receives all the movies relevant to the title he searched for.

Frequently Bought Together

Price For All Three: $50.44

Add all three to Cart   Add all three to Wish List

Some of these items ship sooner than the others. Show details

☑ **This Item:** Ocean's Eleven / Ocean's Twelve / Ocean's Thirteen (Triple Feature) [Blu-ray] ~ George Clooney Blu-ray $13.49
☑ The Last Samurai [Blu-ray] ~ Tom Cruise Blu-ray $6.99
☑ The Dark Knight Trilogy (Batman Begins / The Dark Knight / The Dark Knight Rises) [Blu-ray] ~ Christian Bale Blu-ray $29.96

What Other Items Do Customers Buy After Viewing This Item?

Ocean's Trilogy (Ocean's Eleven / Ocean's Twelve / Ocean's Thirteen) [Blu-ray] ~ Brad Pitt Blu-ray
★★★★★ (118)
$29.60

The Dark Knight Trilogy (Batman Begins / The Dark Knight / The Dark Knight Rises) [Blu-ray] ~ Christian Bale Blu-ray
★★★★★ (43)
$29.96

The Last Samurai ~ Tom Cruise Amazon Instant Video
★★★★★ (790)
$9.99

The Last Samurai [Blu-ray] ~ Tom Cruise Blu-ray
★★★★★ (790)
$6.99

FIGURE 9.7.3 Recommendations and personalization.

FIGURE 9.8 Search and recommendation process.

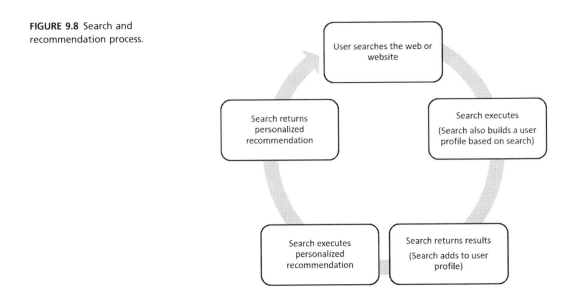

3. John Doe also receives recommendations and personalized offers along with the result sets.

How does the system know what else John Doe will be interested in purchasing and how sure is the confidence score for such a recommendation? This is exactly where we can apply the framework for machine learning as shown in Fig. 9.8.

The first step of the process is a user login or just anonymously executing a search on a website. The search process executes and also simultaneously builds a profile for the user. The search engine produces results that are shared to the user if needed as first pass output, and adds the same to the user profile and as a second step executes the personalized recommendation that provides an optimized search result along with recommendations.

In this entire process after the first step, the rest of the search and recommendation workflow follows the machine learning technique and is implemented with the collaborative filtering and clustering algorithms. The user's search criteria and the basic user coordinates including the website, clickstream activity, and geographical data are all gathered as a user profile data, and are integrated with data from the knowledge repository of similar prior user searches. All of this data is processed with machine learning algorithms and multiple hypothesis results are iterated with confidence scores and the highest score is returned as the closest match to the search. A second pass of the result set and data from knowledge repository is processed to optimize the search and this data is returned as personalized offers to the user. Often sponsors of specific products and services provide such offers with incentives which are presented to the user by the recommender algorithm output.

How does machine learning use metadata and master data? In the search example we discussed, the metadata is derived for the search elements and tagged with additional data as available. This data is compared and processed with the data from the knowledge repository, which includes semantic libraries, and master data catalogs when the machine learning algorithm is executed. The combination of metadata and master data along with use of semantic libraries provides a better quality of data to the machine learning algorithm, which in turn produces better quality of output for use by hypothesis and prediction workflows.

Processing data that is very numeric like sensor data, or financial data or credit card data will be based on patterns of numbers that execute as data inputs. These patterns are processed through several mathematical models and their outputs are stored in the knowledge repository that then shared the stored results back into the processing loop in the machine learning implementation.

Processing data such as images and videos uses conversion techniques to create mathematical datasets for all the nontextual elements. These mathematical datasets are processed through several combinations of data mining and machine learning algorithms including statistical analysis, linear regression, and polynomial curve fitting techniques, to create outputs. These outputs are processed further to create a noise free set of outputs, which can be used for recreating the digital models of images or video data (image only and not audio). Audio is processed as separate feeds and associated with video processing datasets as needed.

Machine-learning techniques reduce the complexity of processing big data. The most common and popular algorithms for machine learning with web-sale data processing are available in the open-source foundation as Apache Mahout project. Mahout is designed to be deployed on Hadoop with minimal configuration efforts and can scale very effectively. While not all machine learning algorithms mandate the need for an enterprise data scientist, this is definitely the most complex area in the processing of large datasets and having a team of data scientists will definitely be useful for any enterprise.

As we see from the discussions in this chapter, processing big data applications is indeed a complex and challenging process. Since the room for error in this type of processing is very minimal if allowed, the quality of the data used for processing needs to be very pristine. This can be accomplished by implementing a data-driven architecture that uses all the enterprise data assets available to create a powerful foundation for analysis and integration of data across the Big Data and the DBMS. This foundational architecture is what defines the next generation of data warehouse, where all types of data are stored and processed to empower the enterprise toward making and executing profitable decisions.

Building the application is the foundational goal and that is what our next chapter of discussion is focused.

Additional reading

The Top 5 Corporate Governance Best Practices That Benefit Every Company (https://www.mcinnescooper.com/publications/legal-update-the-top-5-corporate-governance-best-practices-that-benefit-every-company/).

Effective Business Governance (https://ascelibrary.org/doi/10.1061/%28ASCE%29LM.1943-5630.0000128).

Corporate Governance 2.0 (https://hbr.org/2015/03/corporate-governance-2-0).

DAMA DMBOK – Data Governance (https://www.oreilly.com/library/view/dama-dmbok-data-management/9781634622479/Chapters-5.xhtml).

The Data Governance of Personal (PII) Data (https://www.dataversity.net/dama-slides-the-data-governance-of-personal-pii-data/).

10

Building the big data application

Storyboard development is the first task for developing all applications and associated outcomes, features, transformations, visualization, usage, exploration of data, and other details. The storyboard has to be built based on the following:

- Business use case that has led to the build of the application. The trick here is to not follow the business thought but develop an application flow that will mimic the business thought using storyboard features, data, and logical flow.
- Logical flows that transform data with appropriate business rules and research rules as "applied.iden"
- Log and lineage data collection.
- Requirements for repeatability, pause and proceed, and stop of the flow if needed.
- Conceptual process flows for the steps to be built for the application.
- Visualization requirements and conceptual models of screens as drawn manually.
- Security requirements.
- Fault tolerance and recovery from failure requirements.

This storyboard session will need to be done for each step of the application, which will be useful to determine the programming and development process. There will be features that will be core to the application and used as master library of functionality. These libraries will need to be segmented as core and their application is across multiple touchpoints and this needs to be added to the governance process to prevent chaos later if and when changes are needed to be made and they affect users who did not request for additions or modifications. These are foundational elements that can be defined and processed throughout the lifecycle of the application. Each storyboard will need the following data to be collected and mandated as necessary:

- Storyboard project name
- Storyboard application name
- Storyboard id
- Storyboard date
- Storyboard participants
- Storyboard requirements
- Storyboard data
- Storyboard transformations
- Storyboard compute
- Storyboard error messages
- Storyboard error handling
- Storyboard governance requirements

Building Big Data Applications. https://doi.org/10.1016/B978-0-12-815746-6.00010-7

- Storyboard additions
- Storyboard modifications

In the world driven by big data and internet of things, we need this approach to provide a focus, a starting point and ensure a successful completion of the process. In the earlier chapters where we discussed several examples, the success came as the process optimized, but the early on approach will make a big difference and avoid the classic programmer burnout and failure, which we have seen in the industry many times.

Use cases are the most important piece of the storyboard and the entire application is delivered based on acceptance by the end users. Think of CERN and the experiments to find the Higgs boson particle, if the use case was not specific to what was being experimented, the teams over the years would have lost sight of the project and the applications developed, maintained, and retired through the lifecycle of the experiments would have been failures. The use case needs to be developed with senior leaders who can provide the best in class requirements and also align the outcomes expected from the use case. This is the key to getting the best possible budgets as the use case will be excellent to the enterprise and will be adopted for multiple research projects. An example here is the use case for home stay and study with customers by Procter and Gamble, which resulted in phenomenal success for the research inputs to the "Tide" washing detergent project. If you read "The Game Changer" by A.G.Lafley, he writes in detail the involvement of the executives and the research teams to bring success to the customer, which results in profit for the enterprise. Another research project that can be looked at is the "largest form of cancer and its treatment", yes we are talking about breast cancer, while there are 1462 drugs in the market for treating different issues from the cancer, there are two pharmaceuticals who are on the brink of a good breakthrough, Pfizer and Roche. These giants have a use case for the research and have used the platforms for the data very well to keep focused on outcomes, which have proven to be the tipping point. These use cases will be presented for approvals of the drugs and they are the same as written by the enterprise when they started the process.

Research projects are the backbone for building applications in any enterprise. The old method of creating a roadmap and then delivering business intelligence and analytics are not in vogue. We need to start with defining the outcome needed, for example if the outcome is to read customer sentiment or customer click behavior, we do not need a roadmap but a research project to have formulas to classify the customer based on the outcome and run instant study of the customers as they browse the store on the internet. This research outcome will lead us to a use case, and that will work its way to a storyboard which then leads us to delivering the application. We need to be different in the innovation processing the new world, where we need to experiment and validate, this is why we recommend that all research teams whether in enterprise or university use the data swamp, a raw data layer with dirty data, which is easy to work with and will not interrupt the run of the business process. In drug research or other research programs, this layer is needed to collect the intermediate results for each execution and rerun the process to observe the outcomes in a slower state of study. This will provide two immense benefits, the first is the research, which will validate the need for a better use case, and will deliver instant results for a litmus test observation. In the nest practices approach, this should be the first step taken for any application which will lead to effective storyboards to be created and used by the teams.

Requirements for all applications are driven by the business experts as they have deeper insights into what the enterprise is looking into and wanting to deliver. In the applications world driven with infrastructure like Hadoop or NoSQL, the requirements are needed once the research outcomes are shown by the teams. The reason for this approach is to first isolate the conditions that cause the most impact, and then expand the requirements on those conditions. We learned this lesson in implementing precision medicine initiative, where every patient is specifically isolated and treated by using their own genes which are extracted and mutated in labs, and upon seeing the impact closest to the desired outcome, we inject the mutated gene back into the patient and the outcomes are positive. The requirements were solidified once the outcomes were understood; the same way if the research outcomes are clearly explained to the business experts and executives, their questions will be a better ask. We can stop worrying about that "yet another BI project" finally. The data outcome of the research project will help in managing the requirements with examples and outcomes desired, which is another step to the storyboard.

Log file collection and analysis is another process that building the application in the new world of data mandates. The reason for this requirement is the ability to collect logs as we execute applications is very easy to collect and once the logs are there, running analysis on them will provide the teams a better understanding of what succeeded and what needs to improve. In case of failures, which will be happening, the log analysis will clearly point out where the failure occurred and it will aid in correction of the failure. This is very much the use case for treating patients in the world today where diabetes has become so common and often not treatable with all types of medicines and injections. The doctors have a choice of inserting an IOT sensor that will keep a $24 \times 7 \times 365$ trace of sugars in the patient, and by this observation, they can treat the patient with the right set of options to help them manage the diabetes.

Data requirements will be generated from the research projects and log file analysis. This is a key differentiator in building the big data application, which will provide you the best insights into areas that you are interested in understanding for the betterment of outcomes. This data can be structured, semistructured, unstructured, internal or external, picture, video or email, and can be used once or multiple times. The foundational element is to build the storyboard, you will have the most accurate and precise data that is needed. Metadata and master data associated with this data needs to be documented and added to the storyboard. The reason for this requirement is the same data can be called by different users with different names and it needs to be classified for each team with their specific naming standards.

The next section discusses the actual design, development, and testing of the application. The teams that participated in prior exercises of storyboarding, research, and log file analysis, requirements process are all included and their outcomes of the processes from the prior steps will be added to the next section.

Designing the big data application is a multistep process. We need to do the following in the design:

- The success or failure of a big data project revolves around employees' ability to tinker with information. One challenge is translating a large volume of complex data into simple, actionable business information. "The designer of the application needs to be sure that the application algorithms are sound and that the system is easy to use".
- In the design process, architects and developers will work with scientists to fine-tune complex mathematical formulas. In the foreground is a user, who consumes the outcomes and the insights, which means the application has to filter the data and present it in an easy-to-follow manner so they can probe further. This is a risk area where many a times we fail and do not understand why.
- We need to include user interface designers as key members of the big data application development team. These team members are experts at understanding how end users interact with information and therefore help the design of the interfaces declassifying potential clutter, and present sleek and meaningful interfaces to users.
- Formulas and transformations that have been evolved in the research and analysis segments will need to be documented, and the architects and designers will need to include them into the algorithms and other processes.
- Errors and abends need to be managed in the design process Errors within the application process need separate messages and other errors including data access, storage, and compute errors need separate messages. In any state of error, the recovery from the point of failure should be thought in the design stage. This is really efficient when implemented in the development process as microservices.
- Storage is another area that impacts performance. As datasets become larger, the challenge to process also increases. The current design on the database may partition data, separating older or "almost stale" data from newer information. In the big data infrastructure, the better option is to segment directories divided by ranges of dates or months. The data can be migrated in a weekly maintenance mode and space can be managed with compression and really old data can be archived with metadata as needed.
- Data quality and cleansing takes large time and effort in the database world; this issue is easily solved in the big data world. We can build applications near the raw data layer and let user analyze the data in its raw and dirty layer. This will provide us meaningful data cleansing and other requirements including obfuscation, masking and what data needs to be provided to the end users. The biggest benefit is the applications will perform and deliver insights as designed.
- Metadata, semantic data, and taxonomies will need to be added to the architecture in the design phase. The taxonomies and the rules to process data need to be

identified and aligned with the outcomes. These rules are to be defined by the research team and the experts who know the data and its transformations.

- We will design the application to be delivered on a continuous delivery process, the design benefits that it will deliver includes the following:

  * The primary goal of continuous delivery is to make software deployments pain-less, low-risk releases that can be performed at any time, on demand. By applying patterns such as blue-green deployments it is relatively straightforward to achieve zero-downtime deployments that are undetectable to users.

  * It is not uncommon for the integration and test/fix phase of the traditional phased software delivery lifecycle to consume weeks or even months. When teams work together collaboratively to automate the build and deployment, environment provisioning, and regression testing processes, developers can incorporate integration and regression testing into their daily work and completely remove these phases. We also avoid the large amounts of rework that plague the phased approach. The delivery outcome is **Faster time to market**, which provides better opportunities for market and exposure to prospects and potential customers.

  * When developers utilize automated tools that discover regressions within mi-nutes, teams are freed to focus their effort on user research and higher quality testing activities including exploratory testing, usability testing, and perfor-mance and security testing. By building a deployment pipeline, these activities can be performed continuously throughout the delivery process, ensuring qual-ity is built in to products and services from the beginning.

  * Any successful software product or service will evolve significantly over the course of its lifetime from the perspective of cost. By investing in build, test, deployment, and environment automation, we will substantially reduce the cost of making and delivering incremental changes to software by eliminating fixed costs associated with the release process.

  * **Better products**. Continuous delivery includes a feedback from users throughout the delivery lifecycle based on working software. Techniques such as A/B testing enable us to take a hypothesis-driven approach to product develop-ment whereby we can test ideas with users before building out whole features. This means we will be able to avoid the 2/3 of features we build that deliver zero value to our businesses.

  * Continuous delivery makes releases less painful and reduces team burnout, resulting in better innovation. Furthermore, more frequent releases of software delivery teams provide more engagement with users, learn which ideas work and which do not, and see first-hand the outcomes of the work they have done. By removing the low-value painful activities associated with software delivery, we can focus on what we care about most—continuously delighting our users.

- Storyboard is used in the design phase to ensure that all application requirements and expectations are translated into the technical ecosystem. The technical re-quirements that need to be identified for design include the following:

- Application user interface design
  - Screens
  - Components
  - Appearance
- Application user process flow
  - Starting microservice calls
  - Phased service interfaces and automated calls
  - Passing data between the services and receiving acknowledgments
  - Interface alignment and transformations
- Application data management
  - Data schemas
  - Data files—JSON, XML, PDF, Audio, Video, and Other files
  - Data access rules
  - Data transformations
  - Data write-back into database and files
  - Data error management
  - Data obfuscation and masking rules
- Application computations and transformations
  - Computations in the application
  - Specific process flow and logs
  - Transformations to showcase to user groups
  - Application only transforms
  - Write-back to database and files
- Application security
  - Security for users
  - Security for groups
- Application error management
  - Error messages and management
  - Recovery and process management
  - Fault-tolerance specifics
- Application performance management
  - Performance requirements
  - Optional tuning specifics
  - Optimization guidelines
- Application log management
  - Log files and generation

Development—Testing is to be managed as DevOps process with Kanban methodology. The entire process will drive the agile implementation with several interactions' elements between teams and collaboration of activities including bugs and fixes for the implementation. The agile methodology will be a continuous development and continuous integration process which will be a 4-week release sprint cycle of development, test, and release cycle. The 4-week cycles will require planning to manage several iterations and have integration testing between different teams to happen once the foundational cycle components are developed and delivered. The critical success factors in the development cycle include the following:

- Program plan
- Release plan
- Cycle plan
- Teams with skills for
  - Front end UI/UX
  - Data access and transformation team
  - Library development teams for microservices
  - Testing teams
- Project managers
- Integration testing teams
- End user teams

The next section discusses the Agile approach and Kanban methodology for implementation. The pictures are from Scaled Agile.

Development will be driven by a culture of DevOps, and this culture as defined and implemented with the Safe method includes the following:

- Collaboration and organization—The ability of Agile teams and IT operations teams to collaborate effectively in an ongoing manner, ensuring that solutions are developed and delivered faster and more reliably. This is implemented, in part, by including operations personnel and capabilities on every aspect of the process of development and testing.

- Risk tolerance—DevOps requires a tolerance for failure and rapid recovery, and rewards risk-taking.
- Self-service infrastructures—Infrastructure empowers development and operations to act independently without blocking each other.
- Knowledge sharing—Sharing discoveries, practices, tools, and learning across silos is encouraged.
- Automate everything mindset—DevOps relies heavily on automation to provide speed, consistency, and repeatable processes and environment creation. This is a cool feature to implement in the application development.

Automation is driven by both the need to deliver faster and provide better insights, in the world of internet and "speed of Google". This has led us to adapt to the technique of continuous delivery and release, which is shown in the figure below.

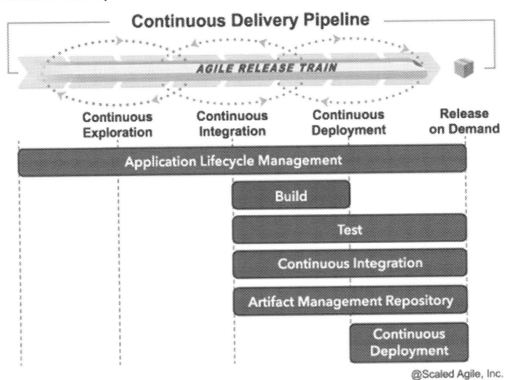

By implementing the continuous delivery pipeline, the automation facilitates faster learning and response to market demand and customer feedback. Builds, testing, deployments, and packaging that are automated improve the reliability of processes that can be made routine. This is accomplished, in part, by building and applying an integrated and automated "tool chain," which typically contains the following categories of tools:

- Application lifecycle management (ALM)—Application and Agile lifecycle management tools create a standardized environment for communication and collaboration between development teams and related groups. Model-based systems engineering provides similar information in many contexts.
- Artifact management repository—These tools provide a software repository for storing and versioning binary files and their associated metadata.
- Build—Build automation is used to script or automate the process of compiling computer source code into binary code.
- Testing—Automated testing tools include unit and acceptance testing, performance testing, and load testing.
- Continuous integration—CI tools automate the process of compiling code into a build after developers have checked their code into a central repository. After the CI server builds the system, it runs unit and integration tests, report results, and typically releases a labeled version of deployable artifacts.
- Continuous deployment—Deployment tools automate application deployments through to the various environments. They facilitate rapid feedback and continuous delivery while providing the required audit trails, versioning, and approval tracking.
- Additional tools—Other DevOps support tools include the following: configuration, logging, management and monitoring, provisioning, source code control, security, code review, and collaboration.

The automation and continuous delivery require us to follow a methodology, the most successful one is Kanban.

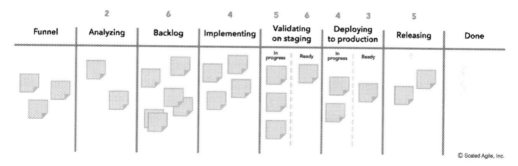

The goals of Kanban include the following:

- Visualize and limit work in process (WIP)—Figure above illustrates an example of a program Kanban board, which makes WIP visible to all stakeholders. This helps teams identify bottlenecks and balance the amount of WIP against the available development and operations capacity, as work is completed when the new feature or functionality is running successfully in production.
- Reduce the batch sizes of work items—The second way to improve flow is to decrease the batch sizes of the work. Small batches go through the system faster

and with less variability, which fosters faster learning and deployment. This typically involves focusing more attention on and increasing investment in, infrastructure and automation. This also reduces the transaction cost of each batch.

- Manage queue lengths—The third way to achieve faster flow is by managing, and generally reducing, queue lengths. For solution development, this means that the longer the queue of work awaiting implementation or deployment, the longer the wait time, no matter how efficiently the team is processing the work. The shorter the queue, the faster the deployment.

Delivery of the application is managed with implementing the Kanban method and aligning with a selection of tools for application development, testing and deployment. Tools include the following and there are more, but some that I have found very useful in the process including development, integration, containers, monitoring, and deployment are as follows:

- **Enov8** is a "DevOps at Scale" platform that sits across your IT landscape that is applications, data, infrastructure, operations and teams, and provides "Enterprise IT Intelligence", delivery-tool integration and holistic command-control that contribute to overall delivery agility.
  - Features:
    - Enterprise level dashboards and reporting
    - DevOps team dashboarding and reporting
    - DevOps environment management (discover, model, and manage)
    - DevOps tool integration
    - DevOps event (and deployment) planning
    - DevOps standardization and collaboration through DevOps Kanban's
    - DevOps task orchestration and automation
    - Lean Service Management and "Self Service" portals.
    - Enov8 is both SaaS and On-premise and provides the necessary framework to deliver enterprise change quickly, safely, and cost effectively.
- **Jenkins is** a DevOps tool for monitoring execution of repeated tasks. It helps to integrate project changes more easily by quickly finding issues.
  - Features:
    - It increases the scale of automation
    - Jenkins requires little maintenance and has built-in GUI tool for easy updates.
    - It offers 400 plugins to support building and testing virtually any project.
    - It is Java-based program ready to run with operating systems like Windows, Mac OS X, and UNIX
    - It supports continuous integration and continuous delivery
    - It can be easily set up and configured via web interface
    - It can distribute tasks across multiple machines thereby increasing concurrency.

- **Vagrant** is a DevOps tool. It allows building and managing virtual machine environments in a single workflow. It offers easy-to-use workflow and focuses on automation. Vagrant lowers development environment setup time and increases production parity.
  - Features:
    - Vagrant integrates with existing configuration management tools like Chef, Puppet, Ansible, and Salt
    - Vagrant works seamlessly on Mac, Linux, and Window OS
    - Create a single file for projects to describe the type of machine and software users want to install
    - It helps DevOps team members to have an ideal development environment
- **PagerDuty** is a DevOps tool that helps businesses to enhance their brand reputation. It is an incident management solution supporting continuous delivery strategy. It also allows DevOps teams to deliver high-performing apps.
  - Features:
    - Provide real-time alerts
    - Reliable and rich alerting facility
    - Event grouping and enrichment
    - Gain visibility into critical systems and applications
    - Easily detect and resolve incidents from development through production
    - It offers real-time collaboration system and user reporting
    - It supports platform extensibility
    - It allows scheduling and automated escalations
    - Full-stack visibility across development and production environments
    - Event intelligence for actionable insights
- **Ganglia is a** DevOps tool that offers teams with cluster and grid monitoring capabilities. This tool is designed for high-performance computing systems like clusters and grids.
  - Features:
    - Free and open-source tool
    - Scalable monitoring system based on a hierarchical design
    - Achieves low per-node overheads for high concurrency
    - It can handle clusters including 2000 nodes
- **Snort** is a very powerful open-source DevOps tool that helps in the detection of intruders. It also highlights malicious attacks against the system. It allows real-time traffic analysis and packet logging.
  - Features:
    - Performs protocol analysis and content searching
    - It allows signature-based detection of attacks by analyzing packets
    - It offers real-time traffic analysis and packet logging
    - Detects buffer overflows, stealth port scans, and OS fingerprinting attempts, etc.

- **Splunk** is a tool to make machine data accessible, usable, and valuable to everyone. It delivers operational intelligence to DevOps teams. It helps companies to be more productive, competitive, and secure.
  - Features:
    - Data drive analytics with actionable insights
    - Next-generation monitoring and analytics solution
    - Delivers a single, unified view of different IT services
    - Extend the Splunk platform with purpose-built solutions for security
- **Nagios** is another useful tool for DevOps. It helps DevOps teams to find, and correct problems with network and infrastructure.
  - Features:
    - Nagios XI helps to monitors components like applications, services, OS, and network protocols
    - It provides complete monitoring of desktop and server operating systems
    - It provides complete monitoring of Java Management Extensions
    - It allows monitoring of all mission-critical infrastructure components on any operating system
    - Its log management tool is industry leading.
    - Network Analyzer helps identify bottlenecks and optimize bandwidth utilization.
    - This tool simplifies the process of searching log data
- **Chef** is a useful DevOps tool for achieving speed, scale, and consistency. It is a Cloud based system. It can be used to ease out complex tasks and perform automation.
  - Features:
    - Accelerate cloud adoption
    - Effectively manage data centers
    - It can manage multiple cloud environments
    - It maintains high availability
- **Sumo Logic is a tool** helps organizations to analyze and make sense of log data. It combines security analytics with integrated threat intelligence for advanced security analytics.
  - Features:
    - Build, run, and secure Azure hybrid applications
    - Cloud-native, machine data analytics service for log management and time series metrics
    - Monitor, secure, troubleshoot cloud applications, and infrastructures
    - It has a power of elastic cloud to scale infinitely
    - Drive business value, growth, and competitive advantage
    - One platform for continuous real-time integration
    - Remove friction from the application lifecycle

- **OverOps** is the DevOps tool that gives root-cause of a bug and informs about server crash to the team. It quickly identifies when and why code breaks in production.
  * Features:
    - Detects production code breaks and delivers the source code
    - Improve staff efficiency by reducing time wasted sifting through logs
    - Offers the complete source code and variable to fix any error
    - Proactively detects when deployment processes face errors
    - It helps DevOps team to spend more time in delivering great features
- **Consul** is a DevOps tool, widely used for discovering and configuring services in any infrastructure. It is a perfect tool for modern, elastic infrastructures as it is useful for the DevOps community.
  * Features:
    - It provides a robust API
    - Applications can easily find the services they should depend upon using DNS or HTTP
    - Make use of the hierarchical key or value store for dynamic configuration
    - Provide supports for multiple data centers
- **Docker** is a DevOps technology suite. It allows DevOps teams to build, ship, and run distributed applications. This tool allows users to assemble apps from components and work collaboratively.
  * Features:
    - Container as a Service (CaaS) Ready platform running with built in orchestration
    - Flexible image management with a private registry to store, manage images and configure image caches
    - Isolates apps in containers to eliminate conflicts for enhancing security
- **Stackify Retrace** is a lightweight DevOps tool. It shows real-time logs, errors queries, and more directly into the workstation. It is an ideal solution for intelligent orchestration for the software-defined data center.
  * Features:
    - Detailed trace of all types of web request
    - Eliminate messy configuration or code changes
    - Provides an instant feedback to check what. NET or Java web apps are doing
    - Allows to find and fix bugs before production
    - Integrated container management with Docker datacenter of all app resources and users in a unified web admin UI
    - Flexible image management with a private registry to store and manage images
    - It provides secure access and configures image caches
    - Secure multi tenancy with granular role-based access control

- Complete security with automatic TLS, integrated secrets management, security scanning, and deployment policy
    - Docker-certified plugins containers provide tested, certified, and supported solutions
- **Monit** is an Open-Source DevOps tool. It is designed for managing and monitoring UNIX systems. It conducts automatic maintenance, repair, and executes meaningful actions in error situations.
    * Features:
        - Executes meaningful causal actions in error situations
        - Helps to monitor daemon processes or similar programs running on localhost
        - It helps to monitor files, directories, and file systems on localhost
        - This DevOps tool allows network connections to various servers
- **Supervisor** is a DevOps tool that allows teams to monitor and control processes on UNIX operating systems. It provides users a single place to start, stop, and monitor all the processes.
    * Features:
        - Supervisor is configured using a simple INI-style config file which is easy to learn
        - This tool provides users a single place to start, stop, and monitor all the processes
        - It uses simple event notification to monitor programs written in any language
        - It is tested and supported on Linux, Mac OS X, FreeBSD, and Solaris, etc.
        - It does not need compiler because it is written entirely in Python
- **Ansible** is a leading DevOps tool. It is a simple way to automate IT for automating entire application lifecycle. It makes it easier for DevOps teams to scale automation and speed up productivity.
    * Features:
        - It is easy to use open-source deploy apps
        - It helps to avoid complexity in the software development process
        - IT automation eliminates repetitive tasks that allow teams to do more strategic work
        - It is an ideal tool to manage complex deployments and speed up development process
- **Code Climate** is a DevOps tool that monitors the health of the code, from the command line to the cloud. It helps users to fix issues easily and allows the team to produce better code.
    * Features:
        - It can easily integrate into any workflow
        - It helps to identify fixes, and improve team's skills to produce maintainable code

- With the code climate, it is easy to increase the code quality
- Allow tracking progress instantly
- **Juju** is an open-source application modeling DevOps tool. It deploys, configures, scales and operates software on public and private clouds. With Juju, it is possible to automate cloud infrastructure and deploy application architectures.
  - Features:
    - DevOps engineers can easily handle configuration, management, maintenance, deployment, and scalability.
    - It offers powerful GUI and command-line interface
    - Deploy services to targeted cloud in seconds
    - Provide detailed logs to resolve issues quickly
- **Scalyr** is a DevOps platform for high-speed server monitoring and log management. Its log aggregator module collects all application, web, process, and system logs
  - Features:
    - Start monitoring and collecting data without need to worry about infrastructure
    - Drop the Scalyr agent on any server
    - It allows to Import logs from Heroku, Amazon RDS, and Amazon CloudWatch, etc.
    - Graphs allow visualizing log data and metrics to show breakdowns and percentiles
    - Centralized log management and server monitoring
    - Watch all the new events arrive in near real-time
    - Search hundreds of GBs/sec across all the servers
    - Just need to click once to switch between logs and graphs
    - Turn complex log data into simple, clear, and highly interactive reports
- **Puppet Enterprise** is a DevOps tool. It allows managing entire infrastructure as code without expanding the size of the team.
  - Features:
    - Puppet enterprise tool eliminates manual work for software delivery process. It helps developer to deliver great software rapidly
    - Model and manage entire environment
    - Intelligent orchestration and visual workflows
    - Real-time context-aware reporting
    - Define and continually enforce infrastructure
    - It inspects and reports on packages running across infrastructure
    - Desired state conflict detection and remediation

The next question that stumps teams is what budgets to ask for application development? What increases of budgets should be planned and how to manage the cycle? Budgets are comprised of total spend cycles for different aspects and these are as follows:

- Applications
  - Requirements
  - Outcomes
  - Use cases
  - Research
  - Validation
  - Performance
  - Security
- Data
  - Systems
  - Access
  - Compute
  - Transformations
  - Overheads
  - Performance
  - Security
- People
  - Program managers
  - Project managers
  - Architects
  - Development leads
  - Development (DevOps) teams
  - QA teams
  - Business analyst teams
  - UAT teams
  - Support teams
- Technologies
  - Commercial vendors
  - Open-source licenses
  - Open-source freeware
- Infrastructure
  - On-premise
  - Cloud
  - Hybrid

The overall cost here is divided into capital spend which is acquisition cost for technology and infrastructure, the innovation spends which includes applications, data and people, the operations spend which includes applications and people. The realistic cost will always have an incremental addition of 10%–20% on the total cost to adjust for inflation and other changes. We should plan the maintenance budget to be 30% of the spend on the overall development budget. This is where the big data platform and the agile methodology will deliver the biggest impact in helping downsize cost models and spend.

The last section is a quick discussion on risks and best practices for application development. Risks are basically there at all times around every aspect of data. The ecosystem of data brings alot of uninvited visitors and intruders, who need to be kept at bay. The risk assessment shown here is a starter kit for all data-related applications, and the areas cover most of the subjects, which can be expanded for your specific requirements.

Risk assessment questions

Information security policy

1. Information security policy document
   * Does an Information security policy exist, which is approved by the management, published and communicated as appropriate to all employees?
   * Does it state the management commitment and set out the organizational approach to managing information security?
2. Review and Evaluation
   * Does the Security policy have an owner, who is responsible for its maintenance and review according to a defined review process?
   * Does the process ensure that a review takes place in response to any changes affecting the basis of the original assessment, example: significant security incidents, new vulnerabilities or changes to organizational or technical structure?

Information security infrastructure

1. Allocation of information security responsibilities
   * Are responsibilities for the protection of individual assets and for carrying out specific security processes clearly defined?
2. Cooperation between organizations
   * Are the appropriate contacts with law enforcement authorities, regulatory bodies, utility providers, information service providers, and telecommunication operators maintained to ensure that appropriate action can be quickly taken and advice obtained, in the event of an incident?
3. Independent review of information security
   * Is the implementation of security policy reviewed independently on regular basis? This is to provide assurance that organizational practices properly reflect the policy, and that it is feasible and effective.

Security of third-party access

1. Identification of risks from third party
   * Are risks from third-party access identified and appropriate security controls implemented?
   * Are the types of accesses identified, classified, and reasons for access justified?

* Are security risks with third-party contractors working onsite identified and appropriate controls implemented?

2. Security requirements in third-party contracts
   * Is there a formal contract containing, or referring to, all the security requirements to ensure compliance with the organization's security policies and standards?

3. Security requirements in outsourcing contracts
   * Are security requirements addressed in the contract with the third party, when the organization has outsourced the management and control of all or some of its information systems, networks, and/or desktop environments?
   * Does contract address how the legal requirements are to be met, how the security of the organization's assets are maintained and tested, and the right of audit, physical security issues, and how the availability of the services is to be maintained in the event of disaster?

Asset classification and control

1. Inventory of assets
   * Is there a maintained inventory or register of the important assets associated with each information system?

Information classification

1. Classification guidelines
   * Is there an information classification scheme or guideline in place; which will assist in determining how the information is to be handled and protected?

2. Information labeling and handling
   * Is there an appropriate set of procedures defined for information labeling and handling in accordance with the classification scheme adopted by the organization?

Security in job definition and resourcing

1. Including security in job responsibilities
   * Are security roles and responsibilities as laid in organization's information security policy documented where appropriate?
   * Does this include general responsibilities for implementing or maintaining security policy as well as specific responsibilities for protection of particular assets, or for extension of particular security processes or activities?

2. Confidentiality agreements
   * Do employees sign confidentiality or nondisclosure agreements as a part of their initial terms and conditions of the employment and annually thereafter?
   * Does this agreement cover the security of the information processing facility and organization assets?

3. Terms and conditions of employment
   * Do the terms and conditions of the employment cover the employee's responsibility for

User training

1. Information security education and training
   * Do all employees of the organization and third-party users (where relevant) receive appropriate information security training and regular updates in organizational policies and procedures?

Responding to security/threat incidents

1. Reporting security/threat incidents
   * Does a formal reporting procedure exist, to report security/threat incidents through appropriate management channels as quickly as possible?
2. Reporting security weaknesses
   * Does a formal reporting procedure or guideline exist for users, to report security weakness in, or threats to, systems or services?

Physical and environmental security

1. Equipment location protection
   * Are items requiring special protection isolated to reduce the general level of protection required?
   * Are controls adopted to minimize risk from potential threats such as theft, fire, explosives, smoke, water, vibration, chemical effects, electrical supply interfaces, electromagnetic radiation, and flood?
2. Power Supplies
   * Is the equipment protected from power failures by using redundant power supplies such as multiple feeds, uninterruptible power supply (ups), backup generator etc.?
3. Equipment Maintenance
   * Is maintenance carried out only by authorized personnel?
   * Is the equipment covered by insurance, and are the insurance requirements satisfied?
4. Securing of equipment offsite
   * Does any equipment usage outside an organization's premises for information processing have to be authorized by the management?
   * Is the security provided for equipment while outside the premises equal to or more than the security provided inside the premises?

5. Secure disposal or reuse of equipment
   * Are storage devices containing sensitive information either physically destroyed or securely over written?
1. Removal of property
   * Can equipment, information, or software be taken offsite without appropriate authorization?
   * Are spot checks or regular audits conducted to detect unauthorized removal of property?
   * Are individuals aware of these types of spot checks or regular audits?

Communications and operations management

1. Documented-operating procedures
   * Does the security policy identify any operating procedures such as Backup, Equipment maintenance etc.?
2. Incident management procedures
   * Does an incident management procedure exist to handle security/threat incidents?
   * Does the procedure address the incident management responsibilities, orderly and quick response to security/threat incidents?
   * Does the procedure address different types of incidents ranging from denial of service to breach of confidentiality etc., and ways to handle them?
   * Are the audit trails and logs relating to the incidents are maintained and proactive action taken in a way that the incident doesn't reoccur?
3. External facilities management
   * Are any of the Information processing facilities managed by an external company or contractor (third party)?
   * Are the risks associated with such management identified in advance, discussed with the third party and appropriate controls incorporated into the contract?
   * Is necessary approval obtained from business and application owners?

Media handling and security

1. Management of removable computer media
   * Does a procedure exist for management of removable computer media such as tapes, disks, cassettes, memory cards, and reports?

Exchange of information and software

1. Information and software exchange agreement
   * Is there any formal or informal agreement between the organizations for exchange of information and software?

Access control
* Does the agreement address the security issues based on the sensitivity of the business information involved?
2. Other forms of information exchange
* Are there are any policies, procedures or controls in place to protect the exchange of information through the use of voice, facsimile and video communication facilities?

Business requirements for access control

1. Access control policy
* Have the business requirements for access control been defined and documented?
* Does the access control policy address the rules and rights for each user or a group of users?
* Are the users and service providers given a clear statement of the business requirement to be met by access controls?

Mobile computing and telecommuting

1. Mobile computing
* Has a formal policy been adopted that considers the risks of working with computing facilities such as notebooks, tablets, and mobile devices etc., especially in unprotected environments?
* Was training arranged for staff that use mobile computing facilities to raise their awareness on the additional risks resulting from this way of working and controls that need to be implemented to mitigate the risks?
2. Telecommuting
* Are there any policies, procedures and/or standards to control telecommuting activities, this should be consistent with organization's security policy?
* Is suitable protection of telecommuting site in place against threats such as theft of equipment, unauthorized disclosure of information etc.?

Business continuity management

Aspects of business continuity management

1. Business continuity management process
* Is there a managed process in place for developing and maintaining business continuity throughout the organization? This might include Organization wide Business continuity plan, regular testing and updating of the plan, formulating and documenting a business continuity strategy etc.,
2. Business continuity and impact analysis
* Are events that could cause interruptions to business process been identified? Example: equipment failure, flood and fire.

- Was a risk assessment conducted to determine impact of such interruptions?
- Was a strategy plan developed based on the risk assessment results to determine an overall approach to business continuity?

3. Writing and implementing continuity plan
   - Were plans developed to restore business operations within the required time frame following an interruption or failure to business process?
   - Is the plan regularly tested and updated?

4. Business continuity planning framework
   - Is there a single framework of Business continuity plan?
   - Is this framework maintained to ensure that all plans are consistent and identify priorities for testing and maintenance?
   - Does this identify conditions for activation and individuals responsible for executing each component of the plan?

5. Testing, maintaining, and reassessing business continuity plan
   - Are the Business continuity plans tested regularly to ensure that they are up to date and effective?
   - Are the Business continuity plans maintained by regular reviews and updated to ensure their continuing effectiveness?
   - Are procedures included within the organizations change management program to ensure that Business continuity matters are appropriately addressed?

If the assessment reveals pitfalls or answers which are doubtful in nature, you can be sure that these areas are the highest risks and they need to be addressed. If these are the risks and pitfalls, what are the best practices? That is a tough question to answer as the areas are quite a few for best practices and here is a brief on the same, and there are a few books that cover these aspects in detail with the topic titled around "Data Driven Applications". In general, there are areas to look for excellence in building applications and they include the following:

- User interface—the biggest impact is to have the right user interface; else adoption will be zero. The best user interfaces come from the users directly and hence engage them to design the screens and components.
- User Expectations—another dark area when it comes to applications is the user expectations of
  - Ease of use
  - Error handling
  - Security
  - Performance
  - Availability
  - Self-learning
  - Ease of platforms adoption

Ensure that these aspects are addressed for the users, and they will be your team for cheers.

* Data—is your best arsenal and armory. Ensure that you have the right permissions, metadata, master data, user driven semantics, and alignment of formats to all screens. This will help you develop and deliver the best possible data layer for applications.
* Skills—are needed to ensure top teams are participating in application development. Ensure that your teams are trained and updated in skills.
* Leadership—is essential to provide success.
* Storyboard—is needed for overall success and engagement.
* Budgets—are needed to keep the lights on.

As we have seen in this chapter of the book, building the application is not trivial and there are several changes that have happened in the world of application development which provide you with so many opportunities. The key mantra here is "fail fast" and you will be very successful if you adopt to this policy. There are some case studies that have been included in the book for you to look and read. The goal of the book is to introduce you to the world of big data applications and data-driven transformation with look into several areas of examples. We will provide more updates on the website over the next few months and welcome specific requests for further examples or case studies.

Additional reading

Storytelling With Data (http://www.storytellingwithdata.com/blog/2015/8/24/how-i-storyboard).

The Ultimate Guide to Storyboarding (https://www.idashboards.com/blog/2018/09/05/the-ultimate-guide-to-storyboarding/).

How to Tell a Story with Data (https://hbr.org/2013/04/how-to-tell-a-story-with-data).

Data Visualization: How to Tell A Story With Data (https://www.forbes.com/sites/nicolemartin1/2018/11/01/data-visualization-how-to-tell-a-story-with-data/#2102c7694368).

11

Data discovery and connectivity

In the world today, data is produced every second and can drive insights that are beyond the imagination of the human threshold, which is funneled by a combination of artificial intelligence, augmented neural networks, backend platform augmented intelligence, streaming data, and machine learning algorithms. We have tipped the threshold of data and analytics processing today with the integration of these tools and technologies. The rate of accessing data as soon as it arrives or even as it is produced to provide insights and impact has evolved into an everyday expectation. This evolution has been even observed by the industry analysts and they are calling it an evolution of business intelligence with augmented intelligence within the solution of business intelligence.

Call it whichever way it makes most sense, data discovery as a process has changed the way we work with data. Imagine if you can identify the data based on a pattern or add newer patterns as they manifest, which makes it easy for analysis prior to integration, we will save 20%—30% of work that is done in the traditional world today. The impact of cloud as a platform is very large and to realize the overall business value, speed is very important, and this platform capability will deliver that requirement. The management of data across the stages of data processing has been changing, and this change requires several steps to be added and modified in the process to ensure constant insights are delivered. These steps have disrupted the traditional processing of data in the enterprise data warehouse, and the disruption is creating a new wave of analytics and insights which are preintegration level insights in many cases and are very useful for predictive and prescriptive analytics which suffered due to lack of availability of data at the right time to create and deliver powerful insights (Fig. 11.1).

Once, centrally administered data warehouses were the most authoritative and comprehensive repository available for analysis of corporate information. However, the data

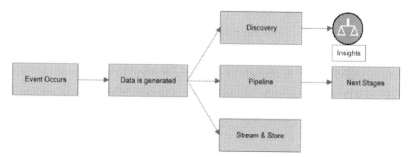

FIGURE 11.1 Data management in the digital age.

Building Big Data Applications. https://doi.org/10.1016/B978-0-12-815746-6.00011-9

199

availability was always having latencies built-in and, in most cases, caused business teams to walk away from a central repository to create their own solutions. This problem has led to silos of data, which I can think of as islands, where the foundation looks similar, but the structure and outcomes are completely different. These systems became legacy over period of time, and several have been discarded and forgotten as we have evolved better tools and access paths. These islands are strewn around all corporations and government agencies, and the issue which concerns us is the cybersecurity attack with hackers trying to penetrate systems on demand and breaching data successfully (Fig. 11.2).

With the passage of time and the evolution of self-service BI tools including Microsoft's Power BI, Tableau and Qlik, people in the business teams have become self-driven to perform analysis work. That motivation has extended a desire to introduce their own data sets and perform similar analysis on them, which have created newer segments of these islands, the only difference being these can be interconnected, and that creates a bigger problem. The issue at heart being how many copies of the same data is prevalent? Who owns them? Who governs them? Who assigns metadata? What is their lifecycle? Are they being removed according to compliance rules? Who audits these silos for assurance of clean data? (Fig. 11.3).

This increased usage patterns of data and analytics has led to growth in volumes and the variety of data sources, which in turn, has led to increased computing and storage requirements and driven increases in cost. The new demands and associated cost increases have drained the legacy data warehouse systems in meeting the required

FIGURE 11.2 The islands.

FIGURE 11.3 The new connected islands of data.

workloads in a cost-effective manner. This is a syndrome that I refer to as "infrastructure fatigue". Even the best in class database systems with all their infrastructure power cannot sustain the workload demands, which drove us to innovate further (Fig. 11.4).

We have evolved in the infrastructure journey with cloud computing, driven by the internet and eCommerce which has quickly changed the landscape of how we can perform computing on data of multiple formats with minimal dependency on databases and moved the users to a world of thrill and immense compute fantasy. However moving infrastructure to the cloud means that we will now use cloud storage, which needs a new data architecture and file design, new governance policies and mechanisms to manage

FIGURE 11.4 Workloads on data—infrastructure fatigue.

the data, and most important of all the need to curb the islands of misfit toys from reoccurring. In this aspect we have evolved the data swamp and data lake layers as the responses to these challenges. The ability to store data in raw format, defer the modeling of it until time of analysis and the compelling economics of cloud storage and distributed file systems has provided answers to manage the problem. The new infrastructure model has evolved quickly and created many offerings for different kinds of enterprises based on size, complexity, usage, and data (Fig. 11.5).

As we evolved the model of computing in the cloud for the enterprise and have successfully adopted the data lake model, the data warehouse is still needed for corporate analytical computes and it has expanded with an addition of the data lake and data swamp layers in the upstream and analytical data hubs downstream. This means we need to manage the data journey from the swamp to the hub, and maintain all lineage, traceability, and transformation logistics, which need to be available on demand.

The multiple layers need to coexist, and our tools and platforms need to be designed and architected to accommodate this heterogeneity. That coexistence is not well accommodated by the tools market, which is largely split along data warehouse-data lake lines. Older, established tools that predate Hadoop and data lakes were designed to work with relational database management systems. Newer tools that grew up in the big data era are more focused on managing individual data files kept in cloud storage systems like Amazon S3 or distributed file systems such as Hadoop's HDFS. The foundational issue is how do we marry the two? (Fig. 11.6).

Enterprises do not want a broken tool chain, they want technologies that can straddle the line and work with platforms on either side of it. They all have multiple sets of data technologies, and the data in each must be leveraged together, to benefit the enterprise. This requires the different databases, data warehouses, data swamps, and other systems with all

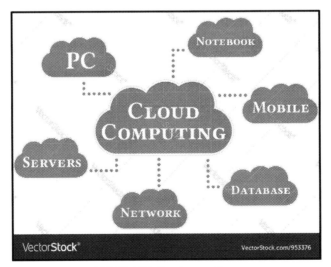

FIGURE 11.5 Cloud computing.

FIGURE 11.6 The new infrastructure divide.

their data sets must coexist in equilibrium, and the data within them must be queried and analyzed in a coordinated fashion. This problem is a challenge for any data architect, and it is very hard to do when companies have separate tools for each broad class of data repository.

The true realization of achieving the harmonized orchestration does not just come from tools that can work with either repository class. The winners are tools that work across both and can bring them together. This is true for both query and analysis: tools that can fetch data from both repository types join that data together and then present visualized insights from that integration, definitely qualify.

The difference between lakes and swamps is much like the distinction between well-organized and disorganized hard disks. Similarly, having a well-organized data catalog, with lots of metadata applied to the data set files within it, makes those data sets more discoverable and the data lake, as a whole, more usable. Beyond just having a good handle on organization, for a catalog to be its best, data from distinct data sources must really coalesce, or the integration at query time will be underserved. Does your data catalog see data in both your data warehouse and data lake? If the answer is yes, can the catalog tell you which data sets, from each, can or should be used together? In other words, does your governance tool chronicle the relationships within and flows between such heterogeneous data sets—do they perform foundational data discovery?

If the answer to all these questions is yes, you are in a good spot, but you are not done yet. Because even if relationships and flows can be documented in the catalog, you then need to determine if this work must be done manually or if the relationships and flows can instead be detected on an automated basis. And even if automatic detection is supported, you need to determine if it will work in instances when there is no schema information that documents the relationships or flows.

Given the data volumes in today's data lakes, both in terms of the numbers and the size of each, discovering such relationships on a manual basis is sufficiently difficult as to be thoroughly impractical. Automation is your savior, and algorithmic detection of such relationships and flows can be achieved through analysis of data values, distribution, formulas, and so forth. But how do we get to this level of sophistication? Who can guide us in this journey?

This is an issue for data management tools, too. When you are going to manage a comprehensive data catalog, then all the data across the enterprise must be in it. The data catalog must be well categorized, tagged, and managed with appropriate metadata, and eventually the exercise should help the enterprise arrest all the silos and integrate the data into one repository with all the interfaces and access mechanisms established.

In establishing the catalogs and managing them, if data sets from the data lake are not properly cataloged, the lake will quickly become mismanaged and lead to even further frustration among users. This is especially the case because of the physical format of a data lake: a collection of files in a folder structure. Similarly, if data sets across the databases are not cataloged in the exercise, they will still be hanging loose and create a mess when the data catalog executes. We need a tool to ensure that all this happens governed and will result in a compliant outcome (Fig. 11.7).

The reason for searching a tool that can be "smart" is primarily to account for external data that comes at us from the world of internet and aligning them with internal corporate data to create meaningful insights, while keeping all rules of compliance intact. These rules of compliance include GDPR, CCPA, Financial rules like Basel III, Safe Harbor Act, and more.

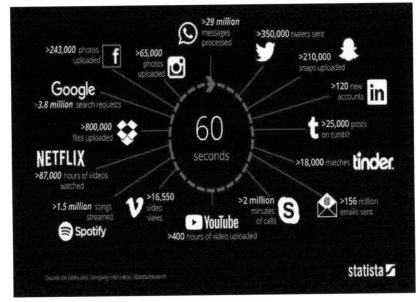

FIGURE 11.7 Data from external sources.

FIGURE 11.8 Frustrated customer; angry executive.

If we do not meet these compliance requirements, we will end up with fines and issues that need to be managed, and the mishandling of data which is out of compliance and can be breached or hacked easily means you will have frustrated consumers and angry executives (Fig. 11.8).

How do we get over this log jam? There are several issues to be answered here

- Catalog of all data
- Catalog with systems info
- Catalog of current data architecture
- Catalog of new data architecture
- Catalog of flexible design

This is where the next generation of technologies comes into play with artificial intelligence and machine learning built into data management architecture. Lots of buzzwords? No, the real-life industry is gravitating toward this point as we need to exceed human boundaries of performance and make it machine driven. Machines can work $24 \times 7 \times 365$ and do not need a break or a holiday. They can learn with Neural Networks and perform tasks which are mundane including data profiling, data formatting, rules engine execution, and more. The delivery of services by these technologies has broken several grounds for data management and even has solved the long-standing desire to have a single platform to manage the data lifecycle from a catalog and metadata perspectives. There are several growth areas as the improvements are done with the solutions and we will see further evolutions for a positive change in the world of data.

There is a lot of hype over artificial intelligence (AI) and machine learning (ML) today than ever before. We have reached a tipping point in terms of the infrastructure, the data processing and analytics ecosystems which is a driver for this hype, which is the next reality that we will undertake across enterprises. Understanding how your company can really make use of them can be a bewildering experience. The industry speak tends to focus on the technical minutiae, making it difficult to see the relevance to your organization or to see your organization as an eligible candidate to adopt and apply AI. The problem generally is not you or your organization, it is the level of the coverage and the focus level of the industry itself. While looking at AI from the perspective of a data scientist is no doubt interesting, for many organizations, doing so is simply getting too far into the weeds.

AI is a combination of frameworks, algorithms, and hyperparameters, and these are all the foundational elements of AI and ML. The underlying goal is the mode of usage and implementation of the ecosystem. You need to decide if you are the driver, passenger or the engineer that will work with the algorithms. To be an engineer, you need to learn and experiment the algorithms, which is a deep dive into the subject. Today most of the end consumers of these technologies just want to drive, and some people only want to be passengers. Using AI in a "driver" or "passenger" capacity is possible; you just need to know where to look. Armed with that knowledge, you can be more intrepid about taking on AI in your own organization.

The passengers on the AI journey can benefit from the power of machine learning without having to learn its rigors, which is deep and wide. That's because many technology products now have AI built-in. For example, Microsoft has added an AI feature to Outlook that sends you a daily message summarizing actions that you promised to take in previous emails. Essentially, such applications capture to-do items from what you wrote without requiring you to enter those as tasks in any explicit way.

If you would like to drive and not just ride, to get closer to the AI technology, yet still not have to immerse yourself in the discipline, the major public cloud providers each offer a number of "cognitive services" to help you do this. For example, Amazon Comprehend can perform sentiment analysis on plain text; Microsoft offers its computer vision service for image classification, activity recognition and more; Google Cloud Video Intelligence extracts metadata from videos; and IBM's Watson Language Translator translates text from one language to another. These cognitive services use machine learning models that are already built and trained by the cloud providers themselves. The gas tank is already full, but would you rather choose your own fuel? You might look for a cloud offering to which customers can bring their own training data. These avoid the need for customers to imagine a predictive scenario, pick algorithms, and tune parameters but still provide a customized offering.

It is good to be a casual driver and get from A to B on your own. But what if you want to drive a stick shift? Going further down, but still staying above the hardcore data science fray, are companies that choose to develop "Auto-ML" systems—something we do at my company but that organizations can pursue independently as well. In such instances, developers (not data scientists) can supply a data set; specify which column within it represents the value they would like to predict and which columns they see as having impactful input on that prediction.

Auto-ML systems can take it from there by selecting the best algorithm and parameter values, then training and testing the machine learning model that results. These systems let developers create their own models without needing to become ML experts.

There are challenges that are to be overcome in this journey with AI. The challenges hover around the same three areas of people, process, and technology. However, the solutions are more deliverable as we have a new generation of engineers who do not worry taking the risk, and our drivers and passengers for the AI segments are bolder and daring to experiment with the new toolkits.

The potential of artificial intelligence is obvious. Adobe Inc has estimated that they can expect 31% of all organizations will start using AI. This statistic is backed up by the fact that there are more startups than ever focusing their operations on AI and the services that it provides to the masses. Roughly 61% of organizations that follow innovative strategies have turned to AI for taking extracts from the data that they previously might have missed. Innovation is an earmark of artificial intelligence, and startup ideas that believe in this ideology can seldom live without the oxygen that is AI.

Not only are marketers confident about AI, but consumers are also starting to grasp its vast potential. About 38% of consumers are beginning to believe that AI will improve customer service. With this growing awareness and popularity, we can expect these numbers to increase further down the line.

Challenges before you start with AI

Organizations are finding it hard to find their footing under the 4 V's of big data; volume, variety, veracity, and velocity. Over 38% of the analytics and data decision-makers from the market reported that their unstructured, semistructured and structured data pools made an increase of 1000 TB in the year 2017. The growth of data is increasing rapidly, as are the initiatives organizations are taking to extract value from it. Herein lie numerous challenges that organizations must overcome to extract full value from AI.

These complications are as follows:

- **Getting Quality Data**
  - Your inference tools would only be as good as the data you have with you. If the data that you are feeding your machines is not structured and flawless, the inference gained from it would barely make the cut for your organization. Thus, the first step of the process is to have quality data.
  - Without the presence of trust in the quality of the data, they would not proceed with their AI initiative. This demonstrates the importance of data quality in AI, and how it changes the perspective of stakeholders involved.
  - The pareto concept applies here, as data scientists are bound to spend almost 80% of their time making data ready for analysis and then the remaining 20% for performing analysis on the prepared data. The creation of these data sets for the ultimate analysis is key to the overall success of the program, which is why scientists have to allot their time.
  - The 80/20 phenomenon has been noted by many analysts online, who believe that 80% of a data scientist's valuable time is spent finding, reorganizing, and cleaning up huge amounts of data.

- **Getting the Best Talent**
  - Once you have quality data you need to understand the importance of recruiting and retaining the best talent in the industry. Since AI is relatively new, the labor market has not matured yet. Thus, you have to be patient in your search for the right talent.
  - Two thirds of the current AI decision makers present within the market struggle with acquiring the right AI talent for their firm. With hiring done, 83% of these companies struggle with retaining their prized employees. The talent shortage obviously goes beyond all technical flaws, as firms need a wide range of expertise to handle AI systems. What is understood here is that traditional recruitment practices are barely implementable, and that organizations need to look for other options.
- **Access to Data**
  - With the increasing rate of data regulations on the horizon, any organization can easily end up on the wrong side of the law if proper measures are not taken. The GDPR or the General Data Protection Regulation by the European Union is one of the most advanced and up-to-date privacy policies for data at the state level. Complying with such policies is mandatory, as noncompliance can leave you in a dire situation.
- **Trust and Data Transparency**
  - There is a trust deficit and the market for AI and analytics is not showing any signs of decreasing over time. While the market has increased and progressed by leaps and bounds, this trust deficit still stands as it is and is not showing any signs of decreasing.

Strategies you can follow to start with AI

With the complications mentioned above, we most definitely will not leave you hanging here. There are certain strategies you can follow for widespread AI implementation. These include the following:

- **Create an AI Vision**
  - Organizations that know what to expect from their AI campaign fare better than those that have no idea about this technology and are just getting involved because their competition has. An AI vision can also act as a list of objectives for the future, so you can tally your end goals with what you planned out before.
- **Build and Manage Customer Journey Centric Teams**
  - The end goal or the mega vision behind AI is to improve customer experience and add value to your offerings. To do this better, you can make customer journey centric teams that follow customers throughout their journey and

improve their experience along the way. Your task goes beyond just making a team, as you will also have to monitor their progress moving forward.

- **Data Accessibility and Culture**
  - ∗ While three fourths of all businesses want to be data driven, only around 29% can agree that they are good at connecting their analytics and data to actively generate insights. If the data you have is not ready for you to get actionable insights, unite your organization around that analysis, and make business decisions based on that.
  - ∗ Data accessibility and culture are necessary for your organization because accessible data enables you to focus on business decisions, move on quickly and build an informed culture where data helps you make better decisions and take better actions.
- **End-to-End AI Lifecycle Management**
  - ∗ End-to-end AI lifecycle management relates to the management of data from its extraction to when it is presented in the form of actionable insight. The process entails different stages like the acquisition, storage, dissemination, learning and implementation of the data. By implementing end-to-end management, you can ensure that your data is always in safe hands.

Most new technologies, when they first emerge, are treated as curiosities. It's the nature of the beast. People are intrigued by technology in general and so, when a new technology comes out, fascination abounds.

But there is a problem with that fascination phenomenon, and there is an issue with putting a new technology on a pedestal. Doing so encourages practitioners and customers alike to romanticize the science of it. Every cool capability, every technical nuance is part of the technology's mystique. While this is cool at the beginning, it inhibits the maturation process as interest in a technology picks up.

In the world of data management, the utility of applied AI is at least double that of the average scenario. That may be a strong statement but consider that data is growing in volume at incredible velocity while its management is being regulated at an ever-growing rate. As the requirements grow (and grow) and the data grows with it, management and governance of that data cannot be done manually. The task is too gargantuan. But the substance of data management involves careful inspection and remediation, so how can it be carried out in any fashion *other* than a manual one?

In fact, AI is built for exactly such a situation: intelligent, judicious examination, and execution, on an automated basis, at scale. Embedded AI is therefore the thing to search for in a data management product, including data governance, data prep, and data discovery. Look for it in these products and check to see how "real" it is. Ask your vendor how they use AI and what functionality is driven by it. The more you know about embedded AI in a product you are considering, the better a purchasing decision you will make.

The AI world's counterpart to a prepared meal is a product with intelligent technology embedded, where machine learning is used behind the scenes to drive or enhance the product's functionality. In this way, customers benefit from AI without having to understand it, or even know it is there. On-staff expertise is no longer required and, certainly, all the rigors of doing machine language work, like algorithm selection and "feature engineering" become the responsibility of the software or service company responsible for the product itself.

That division of labor, where the vendor does the hardest work and the customer consumes the output of that effort, is how technologies become mainstream, have the greatest impact on business, and benefit the most people in the organization. And its advances in *applied* AI where analysts, journalists, customers and, indeed, vendors, and product managers need to focus.

Compliance and regulations

1. Basel Committee on Banking Supervision (BASEL III and BASEL IV)
    1. The Basel Committee rolled out BASEL III, its third set of regulator frameworks around capital and liquidity, in 2010, and is in the process of drafting an updated Basel IV which will likely require higher capital requirements and increased financial disclosure. Basel III and IV share similar goals to Dodd-Frank in that they seek to ensure banks have enough capital on hand to survive significant financial losses, although they differ in the amounts required. The rules establish numerous rules such as Capital-to-Assets Ratio (CAR), Liquidity Coverage Ratio (LCR) and Net Stable Funding Ratio (NSFR) requirements. To meet those requirements, financial service firms again must step up their data reporting and risk management capabilities.
2. Comprehensive Capital Analysis and Review (CCAR)
    1. Spurred by the financial crisis, under the auspices of the Federal Reserve, CCAR mandates certain comprehensive reporting be conducted annually. Effectively, CCAR requires banks to conduct "stress tests" that prove they can "weather the storm" if they were to face the same type of financial challenges experienced during the Great Recession. Banks are then required to report the findings of those tests to regulators.
3. Dodd-Frank Wall Street Reform and Consumer Protection Act
    1. Signed into federal law in 2010, the Dodd-Frank act is a complex piece of legislation passed as a direct response to the financial crisis. Its purpose was to promote "the financial stability of the United States by improving accountability and transparency in the financial system," according to the law's text. Practically speaking, the law implemented standards to limit risk-taking, increase data transparency and improve the efficiency with which data is aggregated and reported to regulators. According to Davis Polk, around 72% of the 390 proposed rules in Dodd-Frank have been met with finalized rules. Rules

have not been proposed to meet 21% of the law's requirements, underscoring that even 7 years later, the regulation's full impact remains uncertain.

4. General Data Protection Regulation (GDPR)

1. When GDPR came into effect on 28 May 2018, it will impose new penalties for companies that run afoul of its cross-border data transfer requirements: fines of up to €20 million ($23.5 million) or 4% of the company's total annual world-wide revenue, whichever is higher. That's just one way in which GDPR seeks to strengthen data protection for EU residents. It puts a greater onus on financial services firms to understand the data they collect and transmit. Importantly, it also impacts banks outside of Europe – any bank with customers in Europe must also comply. Under the regulation, bank customers will need to provide explicit consent for data collection, banks will need to disclose data breaches within 72 h, and banks will need to wipe customers' personal data after a pre-scribed period of time.

5. USA Patriot Act

1. An older and wide-ranging law focused heavily on preventing terrorism, the Patriot Act also includes specific regulatory burdens on financial services com-panies to prevent money laundering, and to report and classify international transactions. Specifically, "suspicious transactions" need to be reported to regu-lators, and banks must identify individuals opening new accounts who meet certain criteria, i.e., owning or controlling 25% or more of a legal entity.

Several of these regulations overlap in terms in their substance and reporting re-quirements – for example, Basel III and Dodd-Frank both seek to increase bank capital and liquidity requirements, even if the method may vary. Each regulation shares the same overall impact, in that they impose significant burden on organizations in how they analyze and report their risk exposure.

The burden flows down to the IT department, which must find ways to collect, aggregate, and understand sensitive corporate data. Speed is important, enterprises have a limited amount of time to find, understand, and report the required information. Even so, they cannot sacrifice data quality, because mistakes in reporting can lead to costly rework or even expensive compliance penalties.

In this world is where we have technologies that have AI built into them and the al-gorithms, they have can be used to implement the data governance needed for both the catalog of data and the compliance requirements. The neural networks that are in the system will recognize data patterns, metadata and provide us with data discovery that will be the first step in creating the data catalog. This data discovery and cataloging is an art of the process of automating data governance, which will ensure the ability to manage data effectively with efficient processes and minimal human intervention providing benefits of error management and correction automatically as much as possible.

One of the vendors in the ecosystem who has a tool with features is IO-Tahoe. I have played with this tool and can confidently tell you that incorporating this tool in the enterprise is the first step toward data democratization and it will bring more collaboration and organization to the process. The tool will discover all the data silos and provide us a map of where all the data is hidden in the enterprise, who are the users of the data, the semantics and usage of the data, the overall status of the system and the process. The information is very effective in managing compliance requirements of GDPR and CCPA both of which work on privacy of information and removal of information from systems and interfaces. The process can also be used to link the multiplicity of the data and create a roadmap to its removal from all systems which are just consumers, thereby creating a single version of the truth. This is more aligned to both master data and metadata functionality, and these systems will be used in the new world of data and on the cloud.

The naysayers of data governance will need to look at these tools and appreciate the process and the optimization provided by the neural network and it is quiet efficiency. The tools are needed for the next generation workforce which are more in the tune of artificial intelligence, which ensures the continuity of business.

Building big data applications is a fantastic exercise, the tools for the exercise will ensure that the build process happens fine and results are delivered as desired.

Use cases from industry vendors
Kinetica

USPS DEPLOYS KINETICA TO OPTIMIZE ITS BUSINESS OPERATIONS

THE BUSINESS

With more than 600,000 employees and a fleet of 215,000 vehicles, the United States Postal Service (USPS) is the single largest logistic entity in the country, moving more individual items in four hours than the combination of UPS, FedEx, and DHL move all year.

In an effort to streamline operations, lawmakers sanctioned the Postal Accountability and Enhancement Act of 2006 to address the technological deficiencies of the current system. The Act called for upgrades to the archaic server and storage platforms, and the adoption of Big Data techniques to improve the ability to respond to market dynamics, enhance competitive practices, and accelerate the delivery process.

Faced with increasingly limited resources and progressively tech-savvy customers, the USPS is tasked with efficiently processing, tracking, and delivering mail on a limited budget.

FIGURE 1: A REGION SHOWING A HEAT-MAP OF DELIVERY POINTS WITH COLLECTION BOX ICONS REPRESENTING MAIL-DROP.

THE CHALLENGE

So how does an organization that makes daily deliveries to more than 154 million addresses using several hundred thousand vehicles and employees create efficiencies based on visual near-real-time data? The United States Postal Service (USPS) turned to Kinetica and its GPU-accelerated database as a first step in improving their ability to improve safety, efficiency, and service without overspending.

Postal customers these days expect sophisticated services like just-in-time supplies, tracking and delivery updates, and dynamic shipment routing. The USPS knew they needed to improve their end-to-end business process performance while reducing costs at the same time. They decided to outfit its entire workforce of USPS carriers with a device that emits its exact geographic location every minute. Armed with this location data, the service aims to improve various aspects of its massive operation, including improving carriers' route efficiency.

To fully optimize data usage, the USPS is rapidly replacing its traditional legacy systems with revolutionary high-performance computing (HPC) solutions. An in-memory relational database was its first choice. But when that technology proved too costly and complex, the USPS looked to Kinetica and

the use of graphical processing units (GPUs). Kinetica is helping the USPS turn their data into knowledge, enabling them to improve efficiency while saving time and money.

For the first time in history, USPS is able to see their entire mobile workforce in real time.

KINETICA SOLUTION

The USPS runs Kinetica on a large cluster composed of 150 to 200 nodes. Each node consists of a single X86 blade server from Hewlett-Packard Enterprise, half a terabyte to a terabyte of RAM, and up to two NVIDIA Tesla K80 GPUs. The system went live in 2014, and was bolstered with a high availability redundancy in November of that year.

With those 200,000+ USPS devices emitting location once every minute, that amounts to more than a quarter billion events captured and analyzed daily, with several times that amount available in a trailing window.

USPS' parallel cluster is able to serve up to 15,000 simultaneous sessions, providing the service's managers and analysts with the capability to instantly analyze their areas of responsibility via dashboards and to query it as if it were a relational database.

And Kinetica has been running since with 5 9's uptime.

RESULTS

Analyzing Breadcrumb Data to Improve Services

The first step in the process was to analyze USPS breadcrumb data. Kinetica was used to collect, process and analyze over 200,000 messages per minute. That data was used to determine actual delivery and collection point locations and enable delivery notifications to mailers and customers. By analyzing this breadcrumb data, the USPS was able to 1) understand where spending would achieve the best results, 2) make faster and more efficient strategic decisions, 3) provide customers with a reliable service, and 4) reduce costs by streamlining deliveries.

Processing Geospatial Data for Real-Time Decision-making

Kinetica was also used to enable visualization of geospatial data (routes, delivery points, and collection point data) so that dispatchers could efficiently plan and graphically view employee territory assignments and take proper action if needed. Kinetica helped the USPS make the best use of routes and to find inefficiencies such as overlapping coverage of assigned areas, uncovered areas, and distribution bottlenecks. They were also able to improve contingency planning if a carrier was unable to deliver to assigned routes and use their work force more efficiency by aggregating point-to-point carrier performance data.

Optimizing Routes

Kinetica springs to life the moment mail carriers depart a USPS Origin Facility. By tracking carrier movements in real time, Kinetica provides the USPS with immediate visibility into the status of deliveries anywhere in the country, along with information on how each route is progressing, how many drivers are on the road, how many deliveries each driver is making, where their last stop was, and more. Course corrections are then made so a carrier is within the optimal geographical boundaries at all times, helping reduce unnecessary transport costs. Optimizing routes results in on-time delivery, fewer trucks handling a greater number of deliveries, and delivery windows narrowed.

FIGURE 2: ZOOMED IN AREA SHOWING COLLECTION BOX METADATA WHEN SELECTED.

FIGURE 3: TERRITORY REASSIGNMENT TOOL SHOWS TWO ROUTE BOUNDARIES AND THE ACTUAL TERRITORY SECTIONS WITHIN THEM THAT CAN BE MOVED BETWEEN THE TWO ROUTES.

IDC INNOVATION EXCELLENCE AWARD RECIPIENT

Due to the success of the project, USPS was named a 2016 recipient of International Data Corporation (IDC)'s HPC Innovation Excellence Award for its use of Kinetica to track the location of employees and individual pieces of mail in real time. "The HPC Innovation Excellence Awards recognize organizations that have excelled in applying advanced supercomputing technologies to accelerate innovation and generate ROI while benefiting science, engineering, and society at large," said Kevin Monroe, senior research analyst at IDC. "USPS' application of Kinetica enhances the quality of service that US citizens receive by giving them a better, more predictable experience sending and receiving mail."

"We're honored to have had the opportunity to partner with the US Postal Service and are humbled by the profound impact our technology is having on their everyday business operations," said Amit Vij, co-founder and CEO, Kinetica. "This IDC Award demonstrates the tremendous value that real-time analytics and visualizations can have when applied to solve supply chains, logistics, and a range of other business challenges."

SUMMARY

The complexities and dynamics of USPS' logistics have reached all-time highs, while consumers have greater demands and more alternative options than ever before. Improving end-to-end business process performance while reducing costs at the same time requires the ability to make fast business decisions based on live data. By implementing Kinetica's GPU-accelerated database, the USPS is expected to save millions of dollars in the years to come, and help them deliver more sophisticated services, achieve more accurate tracking capabilities, ensure safer, on-time deliveries, and increase operational efficiencies.

Lippo group

Getting to Know You: How Lippo Group Leverages a GPU Database to Understand Their Customers and Make Better Business Decisions
Joe Lee, VP, Asia Pacific Japan | January 26, 2018

Lippo Group is a prominent conglomerate with significant investments in digital technologies, education, financial services, healthcare, hospitality, media, IT, telecommunications, real estate, entertainment and retail. Lippo Group has a large global footprint around the world, with properties and services not only in Indonesia, but also in Singapore, Hong Kong, Los Angeles, Shanghai, Myanmar, and Malaysia.

Lippo Group in Indonesia connects and serves over 120 million consumers, aggressively investing in Big Data and analytics technology through OVO. OVO is Lippo Group Digital's concierge platform, integrating mobile payment, loyalty points, and exclusive priority deals.

So, how does one of Asia's largest conglomerates develop 360 degree views of their customers from multiple industry data sources and interact with these customers in a personalized way, while also opening up new revenue streams from data monetization? The answer lies in a combination of Kinetica's centralized digital analytics platform and OVO "Smart Digital Channel" which will help marketers reach out to their audiences and better understand the customer journey.

Challenge: Create a Next-Gen Centralized Analytics Platform

The Lippo Group business ecosystem is vast, with data coming in from multiple lines of business across several industries. The goal was to consolidate fragmented customer data generated by transactional systems from various subsidiaries into a centralized analytics platform. Once consolidated, the data can then be analyzed to generate a 360-degree view of the customer profile and journey.

Solution: Deep and Fast Analytics Powered by Kinetica and NVIDIA

Lippo Group turned to Kinetica and NVIDIA to spur innovation in their big data and analytics strategy. The first project was to a derive insights on customers' digital lifestyles, preference and spending behavior and deliver it in the form of API. Kinetica's distributed, in-memory database combines customer profile, buying behavior, sentiment and shopping trends based on a variety of cross-industry data sources with millisecond latency—powering analytics at the speed of thought.

The new platform is now playing a key role in their multiple industry big data initiative. They are able to develop a full 360-degree profile of each customer, using multiple dimensions of customer attributes to describe their behavior, such as cross channel behavior, CRM/demographic/behavioral/geographic data, brand sentiment, social media behavior, and purchasing behavior via e-commerce, e-wallets, and loyalty programs. Lippo Group can now correlate all customer information and transactions to understand their profile and preferences in order to interact with them in a personalized way. They can also associate household members' activities and preferences as consolidated profiles to deliver family-based personalized offers and improve their service experiences.

By having these types of 360 degree profiles, Lippo Group can improve overall customer experience, conversion rates, campaign take-up rates, inventory/product life cycle management, and future purchasing predictions.

The Technology Behind the Curtain

The underlying technology consists of a Hadoop big data cluster and an analytics layer on Kinetica. In the images below, you can see the evolution in the technology stack that took place to bring Lippo Group's analytics to the next generation:

Before:

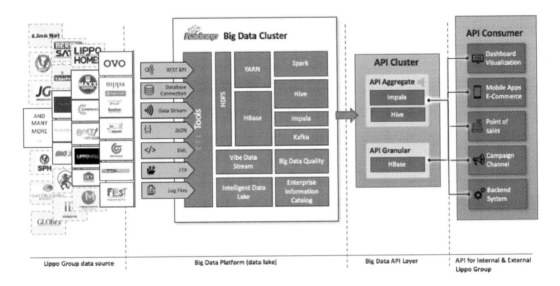

Lippo Group's legacy analytic consisted of Impala, Hive, and HBase

After:

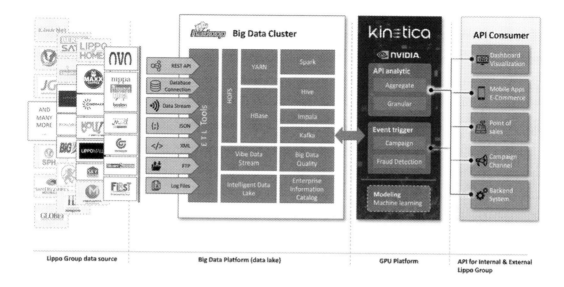

Lippo Group's next-gen analytic API cluster with Kinetica and the ability to augment with machine learning, powered by NVIDIA GPUs

Lippo can leverage any existing digital channel within the organization to deliver personalized messages/offers to their customers and embed machine learning for product recommendations. The same API can be reused multiple times across channels.

OVO will be able to correlate customer, transaction and location information, and geospatial analytics will deliver actionable insights for Lippo business units to better understand business performance and help them outperform competitors. The Big Data API and AI platform enables all OVO touchpoint channels to introduce new and personalized experiences to customers in real time.

Results

Lippo Group's mantra of "deep and fast analytics" is opening up new opportunities for improved customer engagement within the organization's business ecosystem, as well as opening up new revenue streams from data monetization through their interactive digital channel. By having a zbig Data API and AI platform in place, every company within the Lippo Group ecosystem will be able to access the Analytical and Campaign API Marketplace to perform queries through an analytic API that involves huge volumes and rich data sets in subsecond latency.

With the support from Kinetica and NVIDIA, Lippo Group Digital is the first enterprise to integrate an AI, in-memory, GPU database in Indonesia. By harnessing technology and using it with intensity, precision and speed, Lippo Group is now able to understand consumers better, meet their needs and expectations, and make informed operational and product decisions.

Learn more about this project from the Lippo Group presentation at Strata:

Teredata

Case study	Link
Pervasive Data Intelligence Creates a Winning Strategy	https://www.teradata.com/Customers/Larry-H-Miller-Sports-and-Entertainment
Using Data to Enable People, Businesses, and Society to Grow	https://www.teradata.com/Customers/Swedbank
Deep Neural Networks Support Personalized Services for 8M Subscribers in the Czech Republic and Slovakia.	https://www.teradata.com/Customers/O2-Czech-Republic
Operationalizing Insights Supports Multinational Banking Acquisitions	https://www.teradata.com/Customers/ABANCA
More Personal, More Secure Banking. How Pervasive Data Intelligence is Building a More Personalized Banking Experience	https://www.teradata.com/Customers/US-Bank
100 Million Passengers Delivered Personalized Customer Experience	https://www.teradata.com/Customers/Air-France-KLM
Siemens Healthineers Invests in Answers	https://www.teradata.com/Customers/Siemens-Healthineers
Predictive Parcel Delivery, Taking Action in Real Time to Change the Course of Business	https://www.teradata.com/Customers/Yodel

Index

'*Note*: Page numbers followed by "f" indicate figures, "t" indicates tables.'